Natural Gas Hydrates in Flow Assurance

Natural Gas Hydrates in Flow Assurance

Editor

Girish Oak

Natural Gas Hydrates in Flow Assurance

Edited by **Girish Oak**

Printed in 2017

ISBN: 978-1-68117-418-1

Library of Congress Control Number: 2015936536

© 2016 by
SCITUS Academics LLC,
616, Corporate Way, Suite 2, 4766,
Valley Cottage, NY 10989

www.scitusacademics.com

This book contains information obtained from highly regarded resources. Copyright for individual articles remains with the authors as indicated. All chapters are distributed under the terms of the Creative Commons Attribution License, which permits unrestricted use, distribution, and reproduction in any medium, provided the original author and source are credited.

Notice

Reasonable efforts have been made to publish reliable data and views articulated in the chapters are those of the individual contributors, and not necessarily those of the editors or publishers. Editors or publishers are not responsible for the accuracy of the information in the published chapters or consequences of their use. The publisher believes no responsibility for any damage or grievance to the persons or property arising out of the use of any materials, instructions, methods or thoughts in the book. The editors and the publisher have attempted to trace the copyright holders of all material reproduced in this publication and apologize to copyright holders if permission has not been obtained. If any copyright holder has not been acknowledged, please write to us so we may rectify.

Contents

Preface vii

Chapter 1   Numerical Analysis of Thermal Mixing in a Swirler-Embedded Line-Heater for Flow Assurance in Subsea Pipelines 1

Jang Min Park, Dong-Wook Oh, and Jungho Lee

Chapter 2   Experimental Investigations on the Phase Equilibrium of Semiclathrate Hydrates of Carbon Dioxide in TBAB with Small Amount of Surfactant 21

Abhishek Joshi, Jitendra S Sangwai, Kousik Das, and Nagham Amer Sami

Chapter 3   Gas Hydrate Stability and Sampling: The Future as Related to the Phase Diagram 41

E. Dendy Sloan, Carolyn A. Koh, and Amadeu K. Sum

Chapter 4   Molecular Storage of Ozone in a Clathrate Hydrate: An Attempt at Preserving Ozone at High Concentrations 57

Takahiro Nakajima, Taisuke Kudo, Ryo Ohmura, Satoshi Takeya, and Yasuhiko H. Mori

Chapter 5   Towards a Green Hydrate Inhibitor: Imaging Antifreeze Proteins on Clathrates 73

Raimond Gordienko, Hiroshi Ohno, Vinay K. Singh, Zongchao Jia, John A. Ripmeester, and Virginia K. Walker

Chapter 6   Comparison and Analysis of Zinc and Cobalt-Based Systems as Catalytic Entities for the Hydration of Carbon Dioxide 91

Edmond Y. Lau, Sergio E. Wong, Sarah E. Baker, Jane P. Bearinger, Lucas Koziol, Carlos A. Valdez, Joseph H. Satcher Jr, Roger D. Aines, and Felice C. Lightstone

Chapter 7   ESR Study of Interfacial Hydration Layers of Polypeptides in Water-Filled Nanochannels and in Vitrified Bulk Solvents 125

Yei-Chen Lai, Yi-Fan Chen, and Yun-Wei Chiang

Chapter 8	Removal Efficiency of Radioactive Cesium and Iodine Ions by a Flow-Type Apparatus Designed for Electrochemically Reduced Water Production	163
	Takeki Hamasaki, Noboru Nakamichi, Kiichiro Teruya, and Sanetaka Shirahata	

Citations 193

Index 197

Preface

Natural Gas Hydrates in Flow Assurance" provides an expert overview of the practice and theory in natural gas hydrates, with applications primarily in flow assurance. Compact and easy to use, the book provides readers with a wealth of materials which include the key lessons learned in the industry over the last 20 years. Packed with field case studies, the book is designed to provide hands-on training and practice in calculating hydrate phase equilibria and plug dissociation. In addition, readers receive executable programs to calculate hydrate thermodynamics. This title features: case studies of hydrates in flow assurance; the key concepts underlying the practical applications; and, an overview of the state of the art flow assurance industrial developments. Natural gas hydrates are non-stoichiometric, solid substances that consist of a low amount of gas molecules captured in a mesh cage system made up of water molecules.

Editor

Chapter 1

Numerical Analysis of Thermal Mixing in a Swirler-Embedded Line-Heater for Flow Assurance in Subsea Pipelines

Jang Min Park[1], Dong-Wook Oh[2], and Jungho Lee[3]

[1]School of Mechanical Engineering, Yeungnam University, Gyeongsan 712-749, Republic of Korea

[2]Department of Mechanical Engineering, Chosun University, Gwangju 501-759, Republic of Korea

[3]Department of Extreme Thermal Systems, Korea Institute of Machinery and Materials, Daejeon 305-343, Republic of Korea

ABSTRACT

Flow assurance issue in subsea pipelines arises mainly due to hydrate plugs. We present a new line-heater for prevention of hydrate plug formation in subsea pipelines. The line heater has modular compact design where an electrical heater and a swirl generator are embedded inside the housing pipe so that the stream can be heated efficiently and homogeneously. In this paper, flow and heat transfer characteristics of the line heater are investigated numerically, with a particular emphasis on the mixing effect due to the swirl generator.

INTRODUCTION

In offshore plant systems for oil and gas production, flow assurance has arisen as the most crucial issue regarding safety and productivity of the system [1]. Generally the flow in subsea pipeline of the offshore subsea plant system consists of water, oil, natural gas, and sand (or mud), and the term flow assurance refers to ensuring successful flow of the mixture stream through the pipeline. Flow assurance issue in subsea pipelines arises mainly due to gas hydrate plugs which block the pipeline. Hydrate plug can cause serious damage to the system, which is directly related to safety issue, and the production process might be held due to hydrate plugs. Therefore, huge amount of ecomonic cost is involved in removing or preventing hydrate plugs.

Hydrate plug formation is a complex process which depends on hydrodynamic and thermodynamic conditions of the flow, and details of the plug formation mechanism are yet to be investigated. Figure 1 shows a schematic diagram of offshore plant and possible points of hydrate plug formation [1]. In general, the temperature of seabed is around 4 degree Celsius, and thus the flow in subsea pipelines cools rapidly. In this low temperature condition along with high pressure in the pipeline, gas hydrate begins to form on the surface of emulsified water droplet. The hydrate film thickens slowly, converting water droplet into hydrate particle. Wet hydrate particles which have water remaining inside are adhesive with each other and thus they can agglomerate, resulting in hydrate plug. Due to complicate nature of the hydrodynamics and thermodynamics of the mixture with the hydrates, understanding the flow characteristics still remains one of the most challenging issues [2]

Numerical Analysis of Thermal Mixing in a Swirler-Embedded Line...

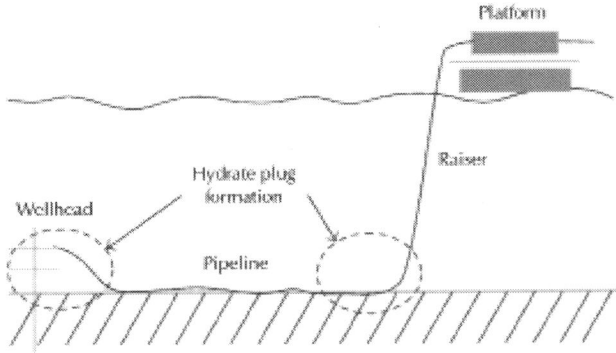

Figure 1: Offshore plant system and possible points of hydrate plug formation in subsea pipeline.

There are several approaches to prevent hydrate plug formation [1, 3–8]. By injecting chemical inhibitors such as methanol and glycol at the well head, hydrate formation can be basically prevented. In this approach, however, the inhibitor should be separated from the stream on the platform, and running costs are relatively high as substantial amount of chemicals are involved. Thermal insulation of the pipeline reduces heat loss of the stream and thus can prevent hydrate formation. Pipe-in-pipe (PiP) and buried pipe are representative examples of this approach and heat transfer characteristics of the buried pipe has been investigated analytically and numerically [7]. In addition, recent experimental investigation has shown that the heat transfer characteristics of the subsea pipeline is significantly affected by the buried depth [8]. Running cost of this approach is considerably lower than that of inhibitor approach. However, this passive approach may not be sufficient for very long pipelines since thermal insulation cannot be complete in practice. For long step-outs of offshore field development, therefore, heat should be supplied to the pipeline by some active manner.

Recently, electrical heating systems for subsea pipeline have been developed by several oil companies [3, 5]. By applying alternating electric current through pipeline, the pipeline can be directly heated due to resistive heating, which is commonly termed as direct electric heating (DEH) method. The pipeline is covered with thermal and electrical insulater, and electric power is supplied via piggyback cable which is mounted on the pipeline (see Figure 2). There is another method called electrically trace heated pipe-in-pipe (ETH-PiP). Basically, ETH-PiP is PiP having electric heater cable coiled around inner pipeline in a helical manner (see Figure 2). This system was recently installed in offshore field for tests and the performance was found to be successfull for industrilization of the system.

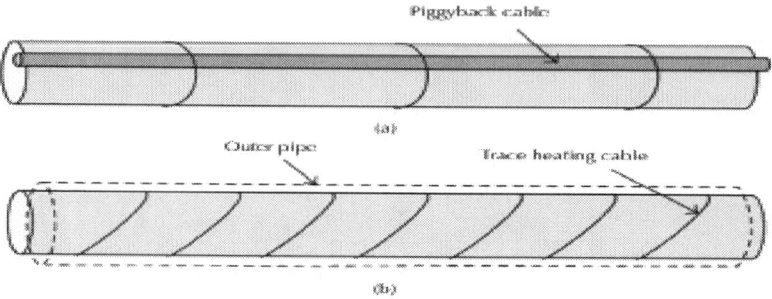

Figure 2: Schematics of direct electric heating method (a) and electrical trace heating pipe-in-pipe method (b).

In this paper, we present a new line-heater for flow assurance in subsea pipelines. Main feature of the line heater is that an electrical heater and a swirl generator are located inside the pipe so that the stream can be heated efficiently and homogeneously. In addition, the line heater has modular compact design and thus it can be installed locally at particular points of possible hydrate plug formation, which would be efficient in terms of the energy consumption and maintenance. We investigate flow and heat transfer characteristics of the line heater numerically in this study. In Section 2, details of the line heater are introduced. Governing equations and numerical method are explained in Section 3 and results are presented in Section 4.

LINE HEATER

In subsea pipelines, there are several points where hydrate plug might be formed as shown in Figure 1. By introducing line heaters locally near these points, hydrate plug formation can be prevented. Figure 3 shows a schematic diagram of the line heater which is installed in between two pipelines via flanges. The line heater consists of a cylindrical cartridge heater, a swirl generator (namely, a swirler), and a housing pipe. The cylindrical cartridge heater is located inside of the swirler, and the lead wires are connected through the L-shape connection.

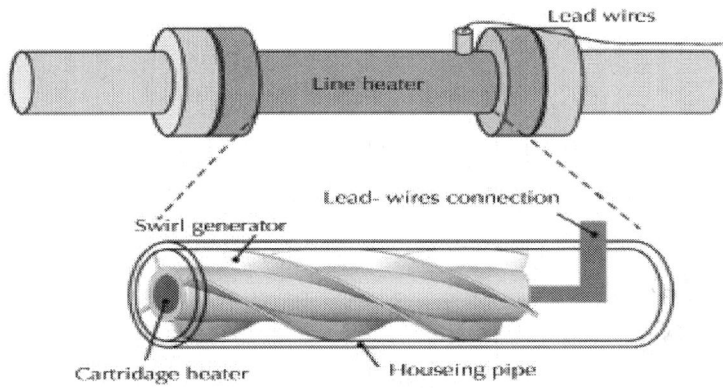

Figure 3: A schematic of line heater installed in pipeline.

The swirler induces three-dimensional complex flow field which enhances heat transfer from the cartridge heater, and thermal mixing of the stream can be significantly improved as well [9, 10]. In addition, hydrate agglomerates already formed upstream can be dispersed and removed by viscous force and applied heat along the line heater.

Details of the line heater are shown in Figure 4. The swirl generator consists of a cylindrical shell and four tapes which are attached on the cylindrical shell with a twisting angle of . The cartridge heater is inserted in a cylindrical cavity inside the swirler. The pipe has inner diameter of D (the pipe thickness is neglected in Figure4). The inner and outer diameter of the cylindrical shell of the swirl generator are c and d, respectively, and the tapes have a thickness of t the cartridge heater has diameter of c.

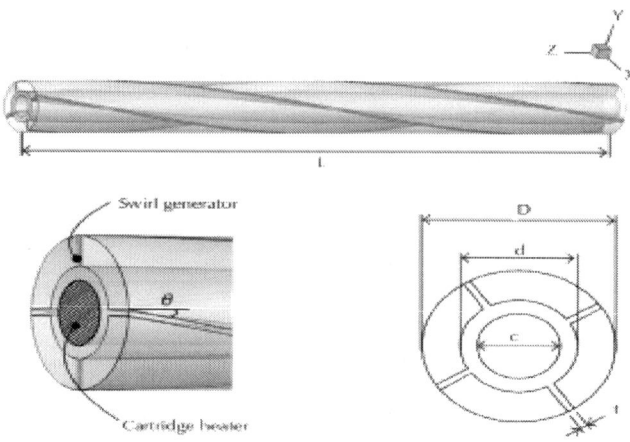

Figure 4: Schematics of a line heater with a swirl number of 0.14.

The geometrical parameters used in this study are L/D=10, d/D=0.6, c/D=0.7, and t/D=0.02. The inlet and outlet are located at z=0, and z=L, respectively. A swirl number S which characterizes the degree of swirl is defined as [11]

$$S = \frac{2}{3}\left(\frac{1-\alpha^3}{1-\alpha^2}\right)\tan\theta, \tag{1}$$

where. $\alpha=d/D$ we study three cases of swirl numbers of 0.14, 0.43, and 0.74 (see Figures 4 and 5). As swirl number increases, the flow will become more complicated and thermal mixing will be enhanced, while the pressure loss increases.

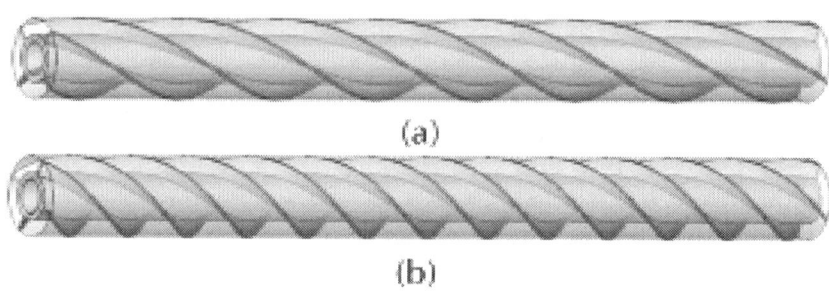

Figure 5: Line heaters with swirl numbers of 0.43 (a) and 0.74 (b).

NUMETRICAL ANALYSIS

Flow and heat transfer characteristics of the line heater are investigated numerically. The cartridge heater is assumed to maintain constant temperature and the housing pipe is assumed to be insulated. The flow is assumed to have uniform constant temperature at inlet ($z/D=0$). As shown in Figure 6, one quater of the swirler and a channel therein are selected as a computational domain. The computational domain consists of two subdomains $\Omega = \Omega_s \ \Omega_f$ where Ω_s represents the solid region and Ω_f represents the fluid region. There is an interface $\Gamma_f = \Omega_s \ \Omega_f$, and five different boundaries of pipe wall Γ_w mid-plane of tapes Γ_s, cartridge heater surface Γ_c, outlet Γ_o, and inlet Γ_i (which is not shown in Figure 6 and is located at the other side of the computational domain). We will solve flow and heat transfer problems in the fluid region Ω_s and heat transfer problem in the solid region. Ω_f The computational domain is discretized by 1575000 hexahedron elements and 1621711 nodes and Figure 6 shows a close-up view of the mesh system. This mesh system is selected after mesh convergence tests which will be discussed later.

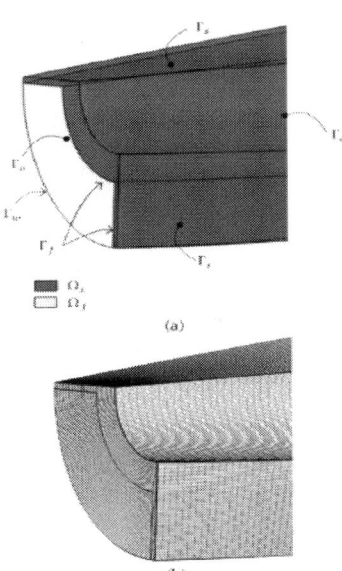

Figure 6: Close-up view of a computational domain (a) and mesh system (b).

We solve the flow and temperature fields at steady state. Governing equations of Navier-Stokes equations and energy conservation equation are written as follows:

$$\nabla \cdot \mathbf{u} = 0 \quad \text{in } \Omega_f,$$
$$\rho \mathbf{u} \cdot \nabla \mathbf{u} = -\nabla p + \nabla \cdot \left[\mu\left(\nabla \mathbf{u} + \nabla \mathbf{u}^\dagger\right)\right] \quad \text{in } \Omega_f,$$
$$\rho \mathbf{u} \cdot \nabla (c_p T) = \nabla \cdot (k_f \nabla T) \quad \text{in } \Omega_f,$$
$$0 = \nabla \cdot (k_s \nabla T) \quad \text{in } \Omega_s, \tag{2}$$

Where u is the velocity, ρ is the density, P is the pressure, μ is the viscosity, T is the temperature, c_p is the heat capacity, and k_f and k_s are the thermal conductivities of fluid and solid, respectively. In this study, material parameters of water and stainless steel are applied for the flow and swirler, respectively.

Boundary conditions are as follows

$$\mathbf{u} = \mathbf{u}_0, \quad T = T_0 \quad \text{on } \Gamma_i,$$
$$p = 0 \quad \text{on } \Gamma_o,$$
$$\mathbf{u} = 0, \quad q = 0 \quad \text{on } \Gamma_w,$$
$$q = 0 \quad \text{on } \Gamma_s,$$
$$\mathbf{u} = 0 \quad \text{on } \Gamma_f,$$
$$T = T_c \quad \text{on } \Gamma_c, \tag{3}$$

Where u_0 is the inlet velocity, T_0 is the inlet temperature, T_c is the cartridge heater temperature, and q is the heat flux. We apply T_0 = 300K and T_c=300 K. It should be noted that the midplane of the paes is assumed to be symmetric plane, and the pipe wall is assumed to be insulated. Boundary names and corresponding boundary conditions are summarized in Table 1.

Table 1: List of boundaries and corresponding boundary conditions

Notation	Boundary name	Boundary conditions
	Inlet	Constant velocity, constant temperature
	Outlet	Zero pressure
	Symmetric	Symmetric condition
	Cartridge heater wall	Constant temperature
	Interface between swirler and fluid	No-slip
	Inner wall of housing pipe	No-slip, insulated

The Reynolds number is defined as

$$\mathrm{Re} = \frac{\rho |\mathbf{u}_0| (D - d)}{2\mu} \tag{4}$$

Which represents the ratio between inertia and viscous forces of the flow. The Péclet number is defined as

$$\mathrm{Pe} = \frac{\rho c_p |\mathbf{u}_0| (D - d)}{2 k_f} \tag{5}$$

Which represents the ratio between advection and diffusion of the heat energy. In this study, the inlet velocity (u_0) is varied from 0.001 m/s to 0.1 m/s so that $10 \leq \mathrm{Re} \leq 1011$ and $70 \leq P_0 \leq 7068$.

Of particular interest of this study is the mixing characteristics in the fluid region. Ω_f. In this regard, species transport problem is solved in the fluid region for mixing visualization purpose [12]. The species concentration C is an artificial quantity for mixing visualization and it does not alter the physical property of the fluid. We solve a simple advection equation of the species concentration C as follows:

$$\frac{\partial c}{\partial t} + \mathbf{u} \cdot \nabla c = 0, \tag{6}$$

Where C varies from zero (black fluid) to unity (white fluid). We introduce a dyed species on the half of the inlet as shown in Figure 7.

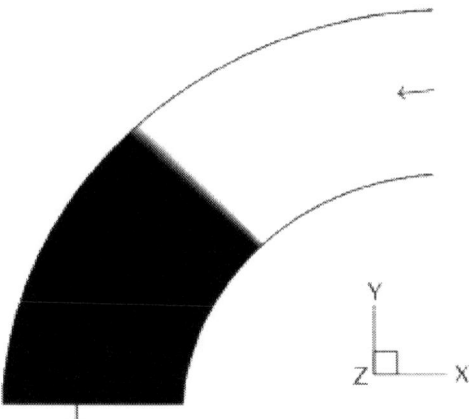

Figure 7: Cross-sectional species distribution at the inlet for mixing characterization. Arrows indicate the twisting direction of the tapes observed from the reference frame.

The degree of mixing is quantified based on a standard deviation σ of the species distribution as follows:

$$\sigma = \sqrt{\frac{1}{N}\sum_{i=1}^{N}(c_i - \bar{c})^2}, \qquad (7)$$

Where N is the total number of data points, c_i is the concentration at point i, and is the mean value of C_i. In a similar manner, we also quantify the thermal mixing based on a standard deviation σ_T of the temperature distribution as follows:

$$\sigma_T = \sqrt{\frac{1}{N}\sum_{i=1}^{N}(T_i - \bar{T})^2}, \qquad (8)$$

Where T_i is the temperature at point I and T is the mean temperature.

All governing equations with the initial and boundary conditions mentioned above are solved by a commercial CFD software ANSYS Fluent which is based on a finite volume method. As mentioned before, the computational domain is discretized by hexahedron elements which are aligned with the boundaries as shown in Figure 6. Mesh convergence test is carried out using four different mesh systems and the results are presented in terms of the standard deviation at the outlet in Table 2. According to the mesh convergence test, mesh system M3 results in an almost converged solution with an appropriate amount of the computational cost. Therefore, mesh system M3 is used for all results in the following.

Table 2: Mesh convergence results in terms of the standard deviation at the outlet. Swirl number = 0.74 and Reynolds number = 1011

Mesh	Number Ele.	
M1	262,500	0.114
M2	393,750	0.182
M3	1,575,000	0.298
M4	3,150,000	0.294
M5	6,300,000	0.285

RESULTS AND DISCUSSIONS

The mixing characteristics are discussed first. Figures 8–10 show cross-sectional species distributions for various Reynolds numbers and swirl numbers along the downstream of the line heater. For the Reynolds number of 10 (see Figure 8), the cross-sectional flow is characterized by a simple rotational flow. One can observe that the cross-sectional configuration is rotating in a clock-wise direction as z/D increases (i.e., along the downstream). This rotational flow becomes more significant as the swirl number increases.

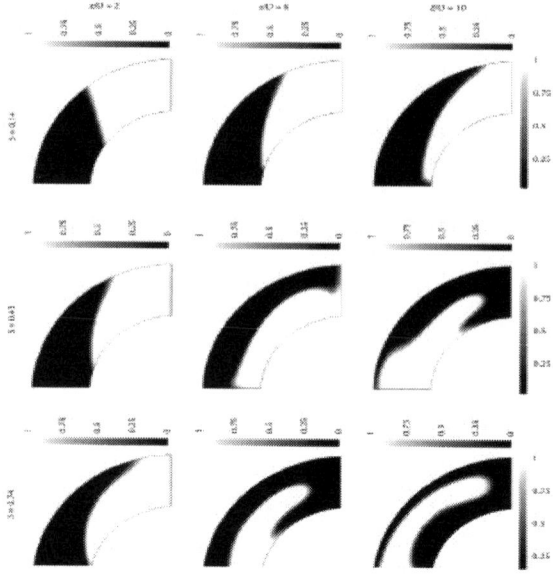

Figure 8: Cross-sectional species distributions for the Reynolds number of 10.

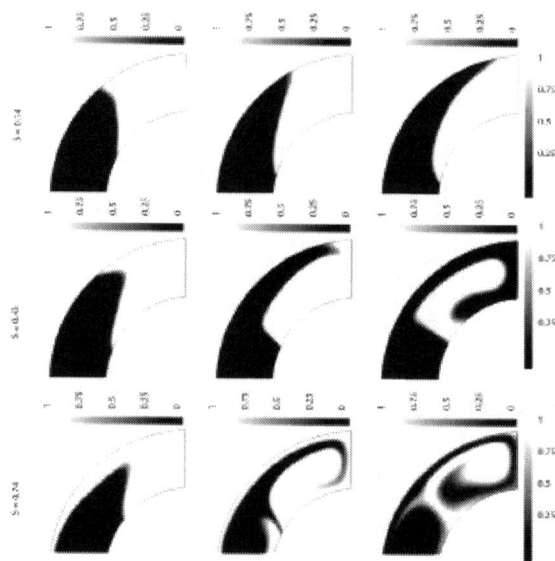

Figure 9: Cross-sectional species distributions for the Reynolds number of 101.

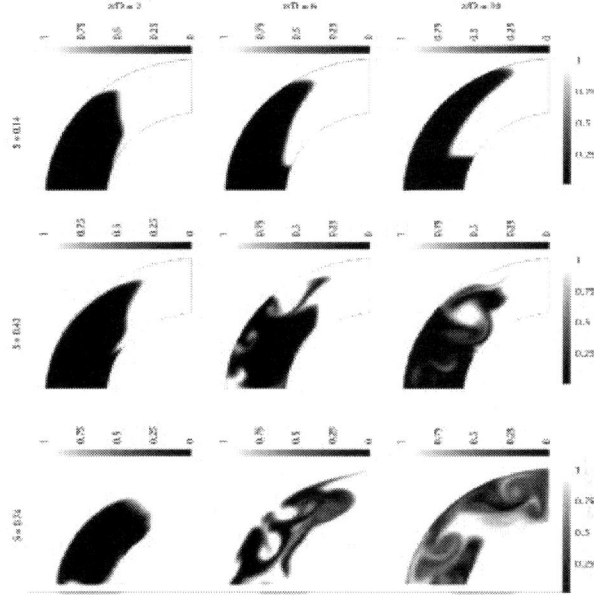

Figure 10: Cross-sectional species distributions for the Reynolds number of 1011.

As the Reynolds number is increased to 101 (see Figure 9), centrifugal force due to rotation of the frame becomes significant particularly for high swirl number case. As shown in the case of swirl number of 0.74, the centrifugal force induces a radial flow, which results in two counter-rotating flows. For a low swirl number of 0.14, this effect is not so significant. Further increase of the Reynolds number to 1011 (see Figure 10) results in a very complicated flow field, especially for swirl number of 0.43 and 0.74 case. The mixing enhancement is notable for the high swirl number of 0.74.

Shown in Figure 11 are mixing quantification results based on ((7)). In Figure 11(a) which is the case for swirl number of 0.74, the standard deviation decreases almost linearly for z/D≥2 , and the decreasing slope becomes larger as the Reynolds number increases. It should be mentioned that the standard deviation remains almost unchanged in the region of z/D ≤2 where the entrance effect is dominant. The standard deviation at the outlet for different swirl numbers is shown in Figure 11(b). For the swirl number of 0.14, the mixing enhancement as the Reynolds number increases is almost negligible.

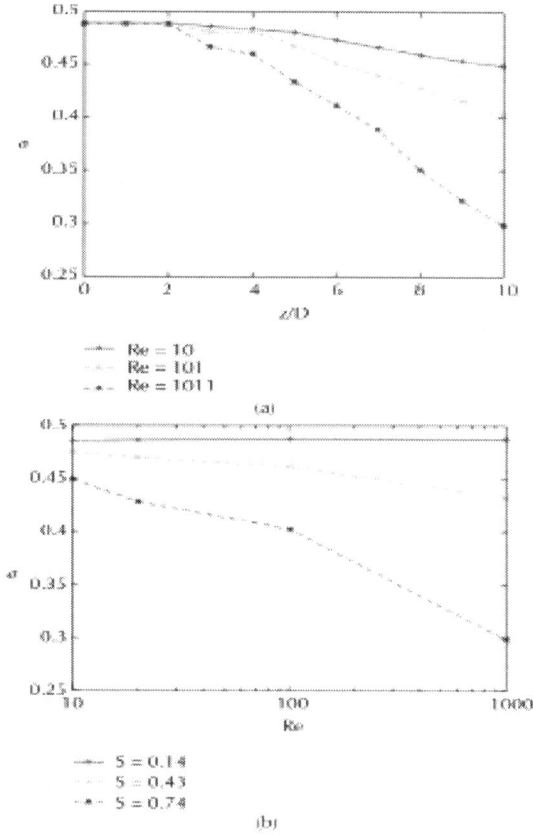

Figure 11: Standard deviation of species distribution (a) along the downstream for swirl number of 0.74 and (b) at the outlet.

Figure 12 shows cross-sectional distribution of the temperature at the outlet of the line heater. At low Re of 10, conduction dominates the heat transfer and thus overall cross-section is heated more uniformly than the other cases of higher Re. The effect of rotational flow as S increases can be also observed. At Re of 101, the flow is heated only around the s wirler when S=0.14. As S is increased, however, one can observe the radial advection enhancing the thermal mixing. At high Re of 1011, axial flow dominates the heat transfer thus most region of the cross-section could not be heated uniformly, though the radial advection effect can be observed as S increases.

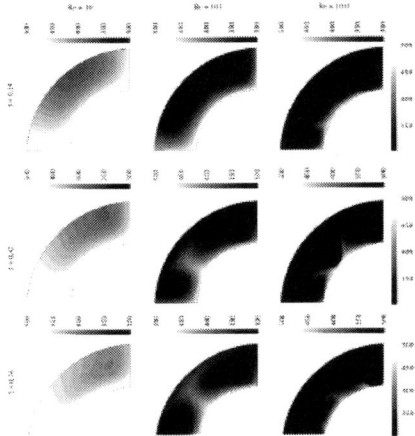

Figure 12: Cross-sectional temperature distributions at the heater outlet.

Thermal mixing of each case is compared quantitatively in terms of the standard deviation (see ((8))) in Figure13. In general, the standard deviation σ_T decreases as the swirl number S increases, which indicates that the swirler enhances the thermal mixing as expected from the mixing analysis results of Figure 11. Relatively low at low Re reflects diffusional mixing by conductive heat transfer, while axial advection at high Re results in decrease of σ_T.

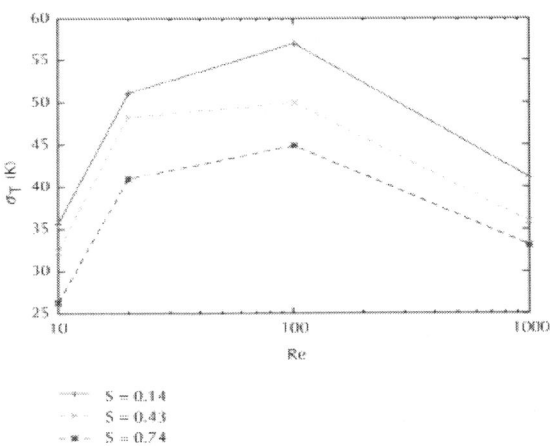

Figure 13: Standard deviation of temperature distribution at the outlet.

Figure 14 shows overall heat transfer coefficient (U) and friction coefficient (f) of the line heater. The overall heat transfer coefficient is defined as

$$U = \frac{1}{A_s(T_c - T_0)} \int_{\Gamma_0 \cup \Gamma_i} \rho c_p T \mathbf{u} \cdot \mathbf{n}\, dA, \qquad (9)$$

Where A_s is wetted surface area of the swirler and n is outward normal vector on the surface. The friction coefficient is defined as

$$f = \Delta P \frac{D_h}{L} \frac{2}{\rho |\mathbf{u}_0|^2}, \qquad (10)$$

Where ΔP is the pressure difference between the inlet and the outlet and D_h is the hydraulic diameter of the channel.

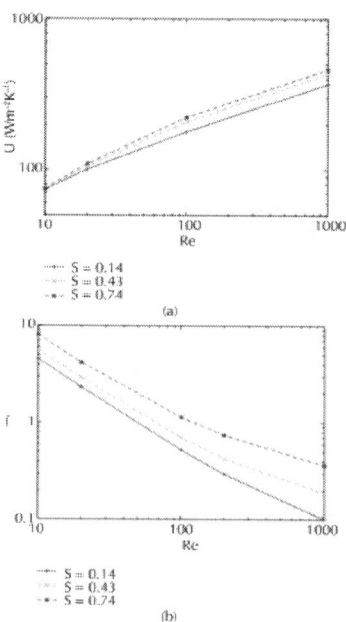

Figure 14: (a) Overall heat transfer coefficient and (b) friction coefficient.

Since the flow is developing flow with the entrance effect involved in, the overall heat transfer coefficient increases with Re in this laminar flow regime. As the swirl number increases, heat transfer is enhanced due to thermal mixing effect especially at Re>20. The friction coefficient is almost inversely proportional to Re for small swirl number of 0.14 since the axial flow is dominant than the swirl effect. As the swirl number increases, the friction coefficient increases, which is especially pronounced at high Re.

CONCLUSIONS

In this paper, we have introduced a new line-heater which can be installed locally at subsea pipelines for prevention of hydrate plug formation. Unlike the previous heating methods, a cartridge heater and a swirl generator are embedded inside the pipe so that the flow can be heated efficiently and homogeneously. To understand the basic flow and heat transfer characteristics of the line heater, we have carried out numerical analysis with a particular emphasis on the thermal mixing effect.

In general, mixing is enhanced as the swirl number and the Reynolds number increase. At low Re$\leq$10, the swirler induces simple rotational flow in the channel cross-section, and the rotational flow becomes more apparent as the swirl number increases. This rotational flow is not effective for mixing, but thermal diffusion results in relatively uniform temperature distribution. At , centrifugal force begins to affect the flow, which induces radial flow and thus enhances the mixing. Further increase of Re to around~1000 results in a very complicated flow field and efficient mixing. As a result, the overall heat transfer coefficient increases with the swirl number and the Reynolds number.

The line heater is expected to be one essential component in subsea pipelines for long step-outs of offshore field development. Although this study is limited to fundamental understanding of the flow and heat transfer characteristics, experimental study in laboratory scale is in progress, which will be presented in a separate publication.

ACKNOWLEDGMENTS

This work was supported by the Industrial Infrastructure Program of the Korea Institute for Advancement of Technology (KIAT) Grant funded by the Korea government Ministry of Trade, Industry and Energy (N009700001).

REFERENCES

1. E. Dendy Sloan, Ed., Natural Gas Hydrates in Flow Assurance, Elsevier, New York, NY, USA, 2011.
2. X. F. Lv, J. Gong, W. Q. Li, B. H. Shi, D. Yu, and H. H. Wu, "Experimental study on natural-gas-hydrate-slurry flow," SPE Journal, vol. 19, no. 2, pp. 206–214, 2014
3. R. C. Fisher, S. Hall, J.-F. Cam, and D. Delaporte, "Field deployment of the world's first electrically trace heated pipe in pipe," in Proceedings of Offshore Technology Conference, Houston, Tex, USA, 2012.
4. O. Heggdal, A. Pedersen, J. K. Lervik, and O.-J. Bjerknes, "Electric heating of pipelines and large export flowlines >30″ and more than 100 km," in Proceedings of the 2012 Offshore Technology Conference, OTC-23646, Houston, Tex, USA, 2012.
5. A. Nysveen, H. Kulbotten, J. K. Lervik, A. H. Børnes, M. Høyer-Hansen, and J. J. Bremnes, "Direct electrical heating of subsea pipelines—technology development and operating experience," IEEE Transactions on Industry Applications, vol. 43, no. 1, pp. 118–129, 2007.
6. M. A. Usman, A. O. Olatunde, T. A. Adeosun, and O. L. Egwuenu, "Hydrate management strategies in subsea oil and gas flowlines at shut-in condition," Petroleum and Coal, vol. 54, no. 3, pp. 191–202, 2012.
7. E. Zakarian, J. Holbeach, and J. E. P. Morgan, "A holistic approach to steady-state heat transfer from partially and fully buried pipelines," in Proceedings of the Offshore Technology Conference (OTC '12), pp. 504–518, Houston, Tex, USA, May 2012.
8. D.-W. Oh, J. M. Park, K. H. Lee, E. Zakarian, and J. Lee, "Effect of buried depth on steady-state heat-transfer characteristics for pipeline-flow assurance," SPE Journal, Article ID SPE-166595-PA, 2014.
9. J. P. Du Plessis and D. G. Kröger, "Heat transfer correlation for thermally developing laminar flow in a smooth tube with a twisted-tape insert," International Journal of Heat and Mass Transfer, vol. 30, no. 3, pp. 509–515, 1987.

10. E. Smithberg and F. Landis, "Friction and forced convection heattransfer characteristics in tubes with twisted tape swirl generators," Journal of Heat Transfer, vol. 86, pp. 39–48, 1964.
11. N. M. Kerr and D. Fraser, "Swirl. Part 1: effect on axisymmetrical turbulent jets," Journal of the Institute of Fuel, vol. 38, pp. 519–526, 1965.
12. I. Manas-Zloczower, Ed., Mixing and Compounding of Polymers: Theory and Practice, Carl Hanser, Munich, Germany, 2009.

Chapter 2

Experimental Investigations on the Phase Equilibrium of Semiclathrate Hydrates of Carbon Dioxide in TBAB with Small Amount of Surfactant

Abhishek Joshi[1], Jitendra S Sangwai[1], Kousik Das1, and Nagham Amer Sami[2,3]

[1]Petroleum Engineering Program, Department of Ocean Engineering, Indian Institute of Technology Madras, Chennai 600 036, India

[2]Department of Petroleum Engineering, College of Engineering, University of Baghdad, Baghdad, Iraq

[3]Department of Chemical Engineering, AC Tech Campus, Anna University, Chennai, 600 025, India

ABSTRACT

Experimental studies are carried out on a semiclathrate hydrate system of carbon dioxide in tetra-*n*-butyl-ammonium bromide (TBAB) with a small amount of surfactant, sodium dodecyl sulfate (SDS), for 5, 10, and 20 wt.% TBAB to determine the phase equilibrium temperature and pressure conditions. It is observed that the presence of SDS did not influence the equilibrium conditions of the semiclathrate hydrate. Re-nucleation (memory) effect of semiclathrate hydrates of CO_2 is studied for few cases of TBAB concentration in an aqueous solution. The equilibrium pressure and temperature conditions obtained for memory effect and regular experimental run without memory effect were observed to be quite close. It is concluded that in the case of no memory effect, with increasing TBAB percentage in the system, the time required for nucleation is reduced. For the same TBAB concentration, the incipient pressure and temperature required for nucleation and re-nucleation of semiclathrate hydrates increase while the time required for re-nucleation decreases.

BACKGROUND

Global warming of the Earth's atmosphere has increased concerns for the conservation of Mother Nature. The melting of glaciers, the encroaching sea line on the coastal areas, decreasing agricultural outputs, health effects on human beings are considered to be some of the major effects of this phenomenon. The burning of fossil fuels and even the use of renewable energies have drastically increased carbon dioxide (CO_2) emission in the atmosphere. It is, therefore, important to look at the proper disposal of carbon dioxide so as to reduce post-use implications of fossil fuels. Carbon capture and sequestration (CCS) is considered to be one of the prime remedy for global warming. However, it is more expensive and technically challenging to have CCS performed in the formation. Additionally, CO_2 leakage from reservoirs may pose a risk to overlying fresh groundwater and also may suddenly increase CO_2 percentage in subsurface environments affecting the marine life [1-3]. In order to mitigate CO_2 leakage and make it more stable in the reservoirs, several methods are being investigated. One of the proposed methods of CO_2 sequestration is using it in the form of gas hydrate and injecting them into the reservoir [4,5]. Additionally, CO_2 can be transported in the form of gas hydrate slurries. However, more efforts have to be made to make the process more economical and safe.

Gas hydrates, also referred to as clathrate hydrates, form under extreme conditions of low temperature (typically <10°C) and high pressure (typically >3 MPa), when gases (guest) like carbon dioxide, methane, nitrogen, come in contact with water (host) [6]. Precise conditions in terms of pressure and temperature depend on the composition of the guest gas. These hydrates get destroyed by destabilizing the phase equilibrium condition, typically by raising the temperature or decreasing the pressure or by employing chemical inhibitors. Semiclathrate hydrates are similar to gas hydrates but have a different lattice structure. This structural difference arises because semiclathrate hydrates are formed when the gas hydrate system contains some thermodynamic promoter, such as tetra-*n*-butyl-ammonium bromide (TBAB), tetra-*n*-butyl-ammonium chloride, tetra-*n*-butyl-ammonium fluoride. TBAB in water forms a semiclathrate hydrate which shares similar physical and structural properties as true clathrate hydrates. The principal difference is that, unlike true clathrates where the guest molecules are not physically bonded within the water structure, in semiclathrate hydrates, the host/guest molecules of TBAB may form the part of the water lattice (host) as well as occupy cages (guest) along with the gas molecules [7-10]. Typically, semiclathrate hydrates are formed at lower conditions of pressure and temperature as compared with the gas hydrate system for the same gas as a guest molecule. Typically, in the semiclathrate structure of TBAB, one TBAB and 38 H_2O molecules form a hydrate structure. Semiclathrate hydrates have a wide range of engineering applications such as carbon dioxide sequestration, transportation and storage of natural gas, and flue gas separation [11-15]. Understating the phase behavior of these hydrate systems forms a precursor for their successful applications.

Arjmandi et al. [7] conducted experiments to determine the equilibrium temperature and pressure of semiclathrate hydrates of methane, carbon dioxide, and nitrogen with varying percentages of TBAB in the aqueous system. The TBAB weight percentage (wt.%) used in this study was in the range of 5 to 42.7 wt.%. A shift in the equilibrium curve of semiclathrate hydrates to the right as compared with that for pure clathrate hydrate systems is observed. This shift implies that semiclathrate hydrates are one of the potential 'thermodynamic' promoters of hydrate formation. The thermodynamic promoter changes the Gibbs energy of hydrate formation which shows implications for the stability condition of hydrates at complex reservoir conditions. Here, the stability of hydrate refers to the equality of fugacity of the hydrate phase and the gas phase. Kinetic promoters also refer to the catalyst which does not affect the phase equilibrium. However, the thermodynamic promoter does affect the phase equilibrium. Several researchers [11-20] conducted experiments to determine the equilibrium conditions of semiclathrate hydrate systems. Duc et al. [11] conducted phase stability experiments on carbon dioxide semiclathrate hydrates for varying TBAB weight percentages from 4.95 to 65 wt.%. Lin et al. [12] studied the phase equilibrium and dissociation enthalpy of carbon dioxide semiclathrate hydrates formed in the presence of TBAB. Sakamoto et al. [13] studied the thermodynamic behavior of hydrogen semiclathrate hydrates formed in an aqueous solution of TBAB. Li et al. [14] conducted experiments for semiclathrate hydrates of carbon dioxide for 5 and 10 wt.% TBAB. Sun et al. [15] determined equilibrium conditions of methane hydrates for 5 to 45 wt.% TBAB in the system. Li et al. [16] and Ding et al. [17] performed experiments to study the formation and dissociation of methane semiclathrate hydrates in an aqueous system containing TBAB. Mohammadi et al. [18,19] conducted experiments to determine the phase behavior of an aqueous system of methane, hydrogen sulfide, carbon dioxide, and nitrogen in the presence of TBAB. Ye and Zhang [20] studied the phase equilibrium of CO_2 in TBAB aqueous solution. The effect of the kinetic promoter (surfactant) such as sodium dodecyl sulfate (SDS) is observed to increase the formation of clathrate hydrates of methane while not affecting the phase equilibria [21]. The use of SDS (kinetic promoter) along with TBAB (thermodynamic promoter) may help to increase the efficiency of the system, making it more useful for CO_2 sequestration and gas storage and transportation. However, the phase behavior of CO_2 hydrates containing both agents is not reported and needs more investigation.

Memory effect has been studied in the crystallization of clathrate hydrates [22]. It is a phenomenon in which the re-nucleation of hydrate crystals occurs at a lower pressure of at least 1 MPa and a higher temperature of at least 2 K than the initial nucleation of hydrates. Parent and Bishnoi [23] conducted experiments to study the memory effect of methane + water system. Takeya et al. [24] performed experiments to analyze memory effect of carbon dioxide + water system. Ohmura et al. [25] observed that the re-nucleation probability and induction time depend on the temperature after the dissociation of initial gas hydrate. Oshima et al. [26] found that the re-nucleation phenomena of semiclathrate hydrates occurred at conditions of temperature 1 K higher than the nucleation temperature of the initial semiclathrate hydrate. It is observed that the effect of the concentration of TBAB on incipient pressure, temperature, and time required for re-nucleation of semiclathrate hydrates has not yet been examined and thus requires investigation.

In this work, an experimental study is performed on the phase behavior of semiclathrate hydrates of carbon dioxide in TBAB aqueous solution with a small amount of surfactant, SDS, for varying weight percentages of TBAB to understand the effect of SDS on the phase stability of the system. Re-nucleation (memory) effect is examined for semiclathrate hydrates of carbon dioxide, and the effect of TBAB concentration on incipient conditions (pressure, temperature, and time) required for re-nucleation is discussed. The results obtained on the phase stability of the combination of CO_2 + TBAB + SDS + H_2O are new and not reported in the literature.

METHODS

A detailed description about the experimental setup is given, followed by discussion on the experimental procedures.

Experimental Setup

The heart of the experimental setup is a high-pressure reactor shown in Figure 1. The volume of the high-pressure reactor is 1 L. The maximum operating pressure of the reactor is 10 MPa. The reactor has a magnetic stirrer which has a maximum rotation speed of 1,000 rpm. The high-pressure reactor is installed with pressure transducers and temperature sensor, Pt-100. The reactor also has a jacket within which ethylene glycol + water solution is circulated from the Julabo® (Julabo Gmbh, Seelbach, Germany) water bath at a desired temperature of about 263.15 K (±1 K). The reactor and the Julabo® water bath is connected to a personal computer (PC; Intel® Core i5, 2 GB RAM; Intel Corporation, Santa Clara, CA, USA) and operated online. The data on pressure and temperature as a function of time are acquired at an interval 30 s and stored in the PC.

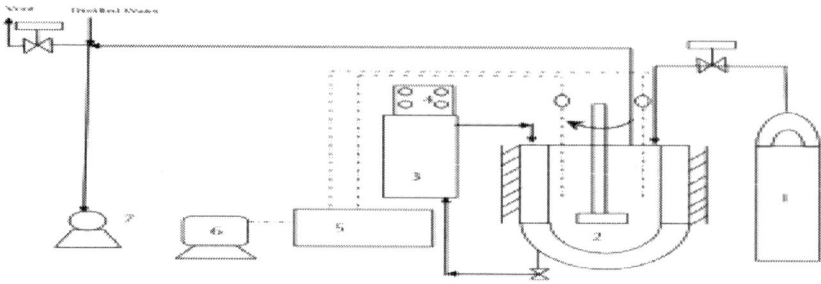

Figure 1: Details of the experimental setup used in this work. (1) CO_2 gas cylinder, (2) high-pressure reactor, (3) Julabo® water bath, (4) temperature control panel, (5) temperature and pressure data acquisition unit, (6) computer, and (7) vacuum pump.

TBAB (in powder form) and SDS used in this work are of ultrapure quality and supplied by Sisco Research Laboratory Private Limited (Mumbai, India). Deionized water obtained from the Millipore® (EMD Millipore, Billerica, MA, USA) deionized setup is used to make the TBAB solution of varying weight percentages as per the requirement. Carbon dioxide gas used in this work is 99.5% pure and is supplied by Bhuruka Gas Agency (Banglore, India). The carbon dioxide cylinder is equipped with a heater at the outlet which helps the smooth flowing of gas to the reactor at the desired pressure.

Experimental Procedure

Typically, two methods, the 'isochoric' and the 'isobaric' methods, are used for the determination of phase equilibrium of semiclathrate hydrates [27-31]. In this work, the isochoric method is used. The temperature of the system is reduced, which results in the reduction of the system pressure. An abrupt fall in the pressure of the system is observed at the onset of hydrate nucleation. After attaining sufficient pressure drop, the system is heated slowly until it attains initial pressure condition. The point of equilibrium is decided when the intersection of tangent is drawn between the heating and cooling lines. The isochoric method is much simpler to conduct and is relatively more accurate as compared to other methods [27, 28].

An aqueous solution of a desired TBAB weight percentage with 0.1 g (0.025 wt.%) of SDS is filled in to the reactor. Carbon dioxide is purged in to the reactor to a pressure of 0.1 to 0.2 MPa for a couple of times to remove dissolved air from the aqueous TBAB solution. The reactor is then pressurized using carbon dioxide and left overnight to allow the system to become stable [31]. The reactor system is then stirred at 480 to 500 rpm and cooled to a lower temperature of about 263.15 K using the Julabo® water bath to form semiclathrate hydrates. As the semiclathrate hydrate forms, the pressure of the reactor falls rapidly. The temperature of the system is then gradually increased at the rate of 2 K/h, followed by an increase at the rate of 0.2 K/h. It is to be noted here that, with the help of several exploratory run *a priori*, we observed a slow heating rate of 0.2 K/h of the reaction mass, which was suitable for our *well-stirred* reactor system to get a *reliable* equilibrium point. To study the re-nucleation phenomena, the system is heated slightly (roughly 0.5 K) above the equilibrium temperature. This was done to ensure that all the gas from the liquid phase in the system has completely escaped to the gas phase. The reactor is then re-cooled to re-form (re-nucleate) the semiclathrate hydrates. Once a semiclathrate hydrate formation is confirmed by sudden drop in the system pressure, the temperature is raised at a rate of 2 K/h followed by 0.2 K/h until the equilibrium point is reached.

RESULTS AND DISCUSSION

In this work, an experimental study is performed on the phase behavior of semiclathrate hydrates of carbon dioxide in TBAB aqueous solution with a small amount of surfactant for varying weight percentages of TBAB. To verify the experimental procedures, experimental investigations were carried out on the pure CO_2 hydrate system for various pressure and temperature conditions to get the phase equilibrium points, followed by studies on semiclathrate hydrates of CO_2 in 5 wt.% TBAB in aqueous solution (without SDS). Figure 1 shows the results obtained for two cases along with literature data. The equilibrium data obtained for the pure CO_2 hydrate system is well matched with those in the literature.

A similar behavior is observed for semiclathrate hydrates of CO_2 in 5 wt.% TBAB aqueous solution, confirming the experimental procedures. As the experimental procedure is verified, several experimental runs were performed to accumulate data sets on equilibrium pressure and temperature conditions of CO_2 semiclathrate hydrate in TBAB aqueous solution with a small amount of SDS for 5, 10, and 20 wt.% TBAB. Table 1 gives the details of experiments performed in this study.

Table 1: Details of the experiments performed in this study

System	TBAB (wt. %)	Data points on equilibriums P and T
CO_2 + TBAB + H_2O	5	6
CO_2 + TBAB + H_2O	10	5
CO_2 + TBAB + H_2O	20	7

SDS weight percentage used in all experimental runs is 0.025 wt. %.

Joshi et al.

Joshi et al. International Journal of Energy and Environmental Engineering 2013 4:11, doi: 10.1186/2251-6832-4-11

Results on Phase Stability

An isochor is defined as the variation of the system pressure with temperature while keeping the volume constant. Each data point on equilibrium temperature and pressure for each TBAB weight percent (as in Table 1) generates separate isochors. A sample isochor is presented in Figure 2 for the experimental run of 20 wt.% TBAB in aqueous solution. The aqueous solution of 20 wt.% TBAB containing 0.025 wt.% SDS and the CO_2 gas in the reactor, at an initial pressure condition as shown by point A (see Figure 2), is cooled at a fast rate (−10 K/h) using cold water from the Julabo® water bath circulated through the jacket of the stirred reactor. Point C represents the onset of nucleation where the system pressure is observed to decrease rapidly, deviating from the normal cooling behavior. Since semiclathrate hydrate formation is an exothermic process, an increase in temperature is observed from points C to D. The cooling is continued until point E, after which the system is heated at the rate of 2 K/h until point G, followed by a rate of 0.2 K/h until the end of the experimental run. At the end of the slow heating process, it is observed that the system retraces its path at point H. The system follows the initial path, B to A, beyond point H. The equilibrium point is decided by the intersection of tangents drawn on the heating line and cooling line near point H as followed by other researchers [27-30]. It is to be noted here that the heating observed under semiclathrate formation can be used to determine heat of formation along with better analytical techniques, such as calorimetry or differential thermal analysis [12,32,33]; however, this measurement does not form the focus of this work.

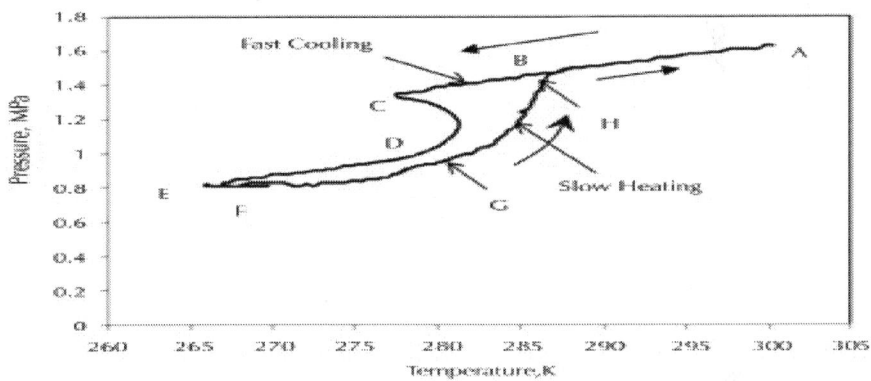

Figure 2: Typical isochor of CO_2 semiclathrate hydrate (for 20 wt.% TBAB in aqueous solution).

Figures 3, 4 and 5 show the experimental data on phase equilibrium for the carbon dioxide semiclathrate hydrate system with 5, 10, and 20 wt.% TBAB with SDS (0.025 wt.%) in aqueous solution, respectively. The equilibrium data obtained in this work is compared with the equilibrium data available in an open literature [7,11-20]. It is found that the data obtained in this work match satisfactorily with the available equilibrium data from the literature for 5 and 10 wt.% TBAB in the system. As mentioned, we have performed experiments with an addition of 0.025 wt.% SDS in TBAB aqueous solution. The obtained results from this work show that the presence of SDS in the system does not influence the phase equilibrium of semiclathrate hydrates of carbon dioxide. The observed results are in accordance with the clathrate hydrate system of methane in the presence of a small amount of SDS [21]. SDS typically helps to increase the rate of formation of the clathrate hydrate system without affecting the equilibrium conditions, so it is termed as 'kinetic' promoters [21]. It is to be noted here that the focus of the present study is to check the effect of the presence of a surfactant on the phase equilibrium of semiclathrate hydrate system. The observed phase equilibrium data points in this study are tabulated in the Table 2.

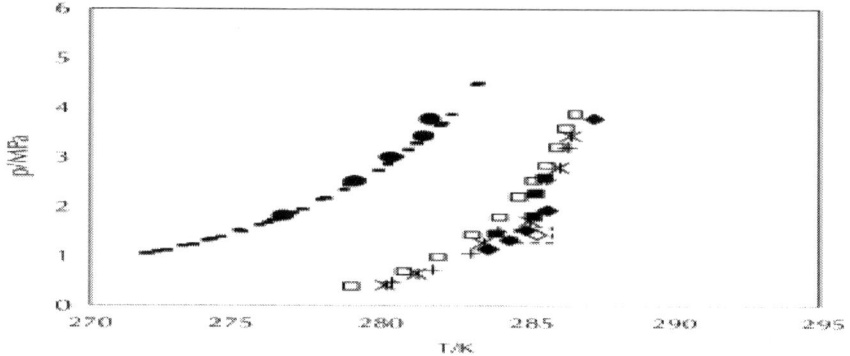

Figure 3: Experimental results for CO_2 hydrate and semiclathrate hydrates with 5 wt.% TBAB. Pure CO_2 hydrate: (black circle) this work and (small black square) Sloan and Koh [6]. Semiclathrate hydrate with 5 wt.% TBAB: (black square) this work; (asterisk) Li et al. [14]; (plus sign) Mohammadi et al. [19]; and (white square) Ye and Zhang [20]. Semiclathrate hydrate with 5 wt.% TBAB + SDS: (black diamond) this work and (white diamond) this work with memory effect.

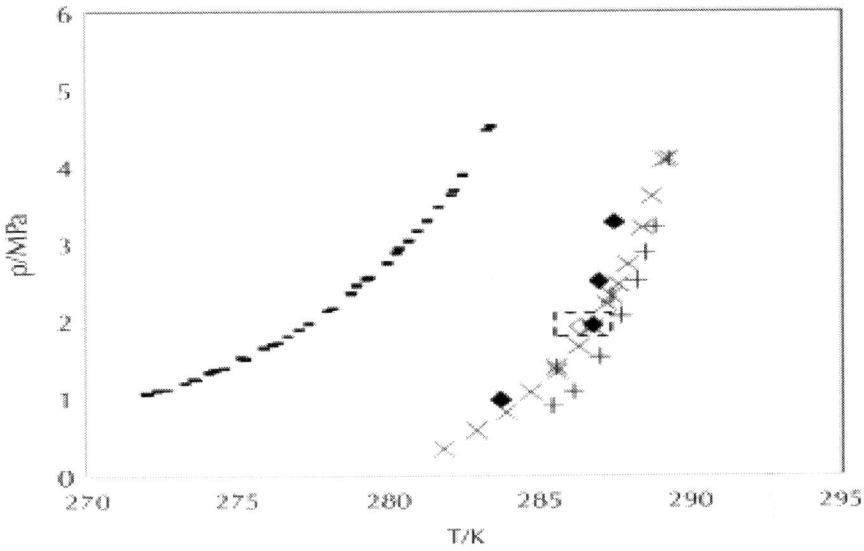

Figure 4: Experimental results for carbon dioxide hydrate and semiclathrate hydrates with 10 wt.% TBAB. Pure CO_2 hydrate: (small black square) Sloan and Koh [6]. Semiclathrate hydrate with 10 wt.% TBAB: (asterisk) Arjmandi et al. [7]; (plus sign) Mohammadi et al. [19]; and (cross sign) Ye and Zhang [20]. Semiclathrate hydrate with 10 wt.% TBAB + SDS: (black diamond) this work and (white diamond) this work with memory effect.

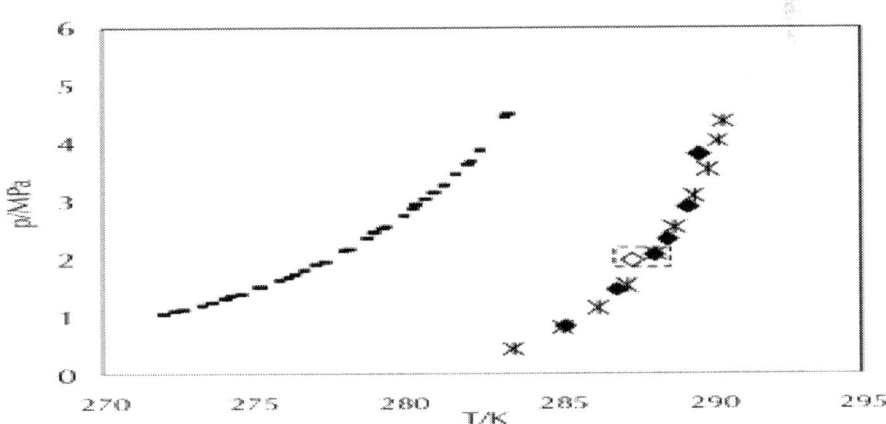

Figure 5: Experimental results for carbon dioxide hydrate and semiclathrate hydrates with 20 wt.% TBAB. Pure CO_2 hydrate: (small black square) Sloan

and Koh [6]. Semiclathrate hydrate with 19 wt.% TBAB: (asterisk) Ye and Zhang [20]. Semiclathrate hydrate with 20 wt.% TBAB + SDS: (black diamond) this work and (white diamond) this work with memory effect.

Table 2: Equilibrium data for various TBAB percentages for semiclathrate hydrates of CO_2

TBAB (wt.%)	Equilibrium temperature (K)	Equilibrium pressure (MPa)
5	285.25	0.84
	286.90	1.47
	287.47	1.96
	288.18	2.07
	288.60	2.33
	289.29	2.89
10	283.68	0.97
	285.79	1.92
	286.73	1.94
	286.93	2.49
	287.46	3.27
20	285.25	0.87
	286.90	1.52
	287.47	2.03
	288.19	2.14
	288.60	2.41
	289.29	2.98
	289.66	3.91

Joshi et al.

Joshi et al. International Journal of Energy and Environmental Engineering 2013 4:11, doi:10.1186/2251-6832-4-11

Re-Nucleation Effect in Semiclathrate Hydrate

The purpose of this study is to get an inference of the effect of the presence of TBAB concentration on the re-nucleation effect in the semiclathrate hydrate system, not to analyze the re-nucleation effect in detail. To understand the memory effect in detail, one may require more sophisticated analytical/visual techniques. A sample isochor for the re-nucleation effect for semiclathrate hydrate of CO_2 is shown in Figure 6 for a case of 20 wt.% TBAB in the system. It is to be noted that the isochor here (Figure 6) is separate from the isochor presented in Figure 2, which is representing another equilibrium point. The isochor with dashed lines in Figure 6 is for initial experimental runs (without memory effect) while the solid line represents the re-nucleation (memory effect) run. The notations for points A, B, and C are the same as those in Figure 2. A', B', and C' represent the points for experimental run with re-nucleation effect. It is to be noted that the formation of semiclathrate hydrate is achieved first and then is dissociated to reach up to equilibrium point B as in Figure 6. After reaching point B, the system temperature is raised by about 0.5 K to ensure that all gas molecules from semiclathrate hydrate crystals have escaped to the bulk gas phase. This is validated by the fact that the P to T line traces back to its original cooling curve after point B (path B to A). This is then followed by re-cooling of the system to form the semiclathrate hydrates again. An analysis is done to see an effect of the TBAB concentration on nucleation condition from several cases of TBAB concentrations studied in this work. For this, out of several experimental runs (as in Table 1), three cases, one each from 5, 10, and 20 wt.% TBAB concentrations, are considered. In order to draw realistic and practical conclusions and inferences, the same initial conditions of system pressure and temperature are considered for the three cases. The three cases, thus, considered for this examination have initial conditions (point A in Figure 6) of system pressure and temperature that are nearly same. This is around 2.25 MPa of initial system pressure and nearly in the range 298.15 to 300.15 K of initial temperature. Table 3 gives a comparative study between the difference in insipient temperature and pressure conditions at which nucleation starts (point C in Figure 6) in the case of the first run and a run with memory effect (second experimental run) along with the corresponding time of re-nucleation of the semiclathrate hydrates in the system for 5, 10 and 20 wt.% TBAB. The equilibrium point obtained for memory effect and the regular experimental run are observed to be quite close (also shown in Figures 3, 4 and 5). It is observed from Table 3 that for the case of no memory effect with increase in TBAB percentage in the system, the time required for nucleation reduces and so is for the case of memory effect. Additionally, for the same TBAB concentration, the incipient pressure and temperature required for nucleation and re-nucleation of semiclathrate hydrates increase, while the time required for re-nucleation decreases. This study, in general, indicates that with the increase of TBAB concentration in the system, it does help to form semiclathrate hydrates at an early stage.

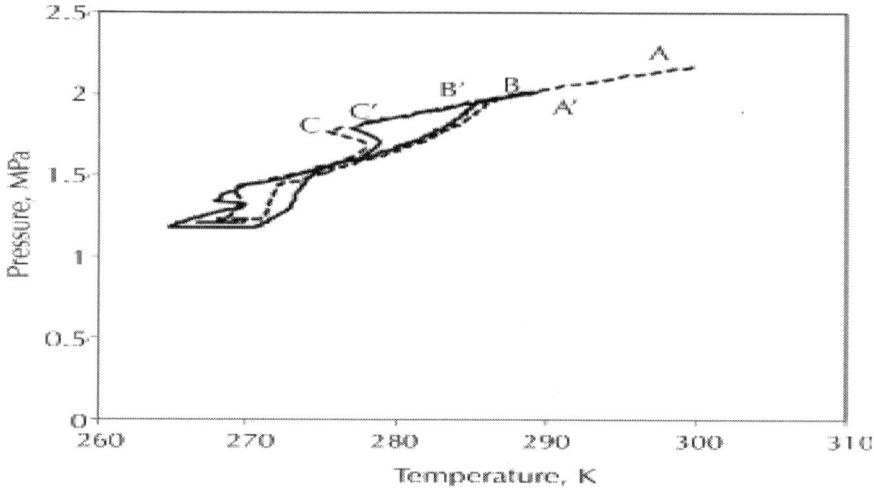

Figure 6: Re-nucleation (memory) effect for semiclathrate hydrates of CO_2 for sample case of 20 wt.% TBAB. Isochor for first run (dashed lines); isochors for memory effect (solid line). The notations on points A, B, and C are the same as those in Figure 2. A ', B', and C' represent experimental points for re-nucleation.

Table 3: Temperature and pressure variation with time during the memory effect phenomenon

Number	TBAB (wt.%)	P and T condition at first run			P and T condition at re-nucleation (memory effect)		
		T_1^a (K)	P_1^a (MPa)	t_1^a (s)	T_2^b (K)	P_2^b (MPa)	t_2^b (s)
1	5	275.5	1.97	510	277.05	2.06	450
2	10	276.65	1.88	420	278.57	1.97	300
3	20	277.15	1.87	330	279.85	1.89	240

P, pressure; T, temperature. $^a T_1$, P_1, t_1 = incipient temperature, pressure, and time at nucleation of semiclathrate hydrates (at point C in Figure 6); $^{b}T_2$, P_2, t_2 = incipient temperature, pressure, and time at nucleation of semiclathrate hydrates (at point C' in Figure 6).

Joshi et al.

Joshi et al. *International Journal of Energy and Environmental Engineering* 2013 4:11, doi:10.1186/2251-6832-4-11

CONCLUSIONS

The experimental study on the phase behavior of semiclathrate hydrate of carbon dioxide in TBAB aqueous solution with a small amount of surfactant for 5, 10, and 20 wt.% TBAB are discussed. The data on phase equilibria of semiclathrate hydrates for 5 and 10 wt.% TBAB are in good agreement with the published data from the open literature. The presence of the surfactant, SDS, did not influence the equilibrium conditions for the semiclathrate hydrate system. The memory effect phenomenon is studied for few cases of TBAB concentration. The increase in the concentration of TBAB in the system helps to reduce the time required for re-nucleation of semiclathrate hydrates. Additionally, for the same TBAB concentration, the incipient pressure and temperature required for nucleation and re-nucleation of semiclathrate hydrates increase, while the time required for re-nucleation decreases.

AUTHORS' CONTRIBUTIONS

AJ developed the experimental setup and carried out the experimental studies. JSS initiated, guided and supervised the work, and drafted the manuscript. NAS and KD carried out validation studies and helped in literature search. All authors read and approved the final manuscript.

AUTHORS' INFORMATION

AJ is currently working as a field engineer in Schlumberger. He finished his M.Tech. in Petroleum Engineering from Indian Institute of Technology Madras (IITM). JSS is working as an assistant professor at IIT Madras. His areas of research interest are in gas hydrates, flow assurance, and enhanced oil recovery. NAS is a student at Anna University and is also a lecturer at the University of Baghdad, Iraq. KD is a project officer working at IIT Madras and is under the M. Tech. program in Chemical Engineering in IIT Madras.

ACKNOWLEDGMENTS

The authors would like to thank the director of the National Institute of Ocean Technology (NIOT) and NIOT-IITM cell for the encouragement towards setting up the laboratory facilities. Thanks to Dr. S Ramesh and Dr. Ramadass from NIOT and Professor SK Bhattacharyya of IIT Madras for their valuable comments and extended cooperation during the course of our work. Financial support from NIOT, Chennai, India, through grant OEC/10-11/105/NIOT/JITE is also gratefully acknowledged.

REFERENCES

1. Hawkes, C: Geomechanical factors affecting geological storage of CO_2 in depleted oil and gas reservoirs. J Can Petro Technol. 44(10), (2005).
2. Little, MG: Jackson, RB: Potential impacts of leakage from deep CO_2 geosequestration on overlying freshwater aquifers. Environ Sci Technol. 44(23), 9225–9232 (2010
3. Zappa, P, Schreiber, A, Marx, J, Haines, M, Hake, JF, Gale, J: Overall environmental impacts of CCS technologies—a life cycle approach. Inter J Greenh Gas Contr. 8, 12–21 (2012)
4. Holder, G, Cugini, A, Warzinski, R: Modeling clathrate hydrate formation during carbon dioxide injection into the ocean. Environ Sci Technol. 29, 276–278 (1995).
5. Qanbari, F, Pooladi-Darvish, M, Tabatabaie, S: CO_2 disposal as hydrate in ocean sediments. J Natu Gas Sci Eng. 8, 139–149 (2012)
6. Sloan, ED, Koh, CA: Clathrate Hydrates of Natural Gases, New York: CRC Press (2008)
7. Arjmandi, M, Chapoy, A, Tohidi, B: Equilibrium data of hydrogen, methane, nitrogen, carbon dioxide, and natural gas

in semiclathrate hydrates of tetra-*n*-butyl ammonium bromide. J Chem Eng Data. 52, 2153–2158 (2007).
8. Shimada, W, Ebinuma, T, Oyama, H, Kamata, Y, Takeya, S, Uchida, T, Nagao, J, Narita, H: Separation of gas molecule using tetra-*n*-butyl ammonium bromide semi-clathrate hydrates crystals. Jap J App Phys Part 2 Lett. 42, L129–L131 (2003).
9. Shimada, W, Ebinuma, T, Oyama, H, Kamata, Y, Narita, H: Free-growth forms and growth kinetics of tetra-*n*-butyl ammonium bromide semi-clathrate hydrate crystals. J Crys Growth. 274, 246–250 (2005).
10. Shimada, W, Shiro, M, Kondo, H, Takeya, S, Oyama, H, Ebinuma, T, Narita, H: Tetra-*n*-butylammonium bromide-water (1/38). Acta Crystall Sect C. 61, 65–66 (2005)
11. Duc, NG, Chauvy, F, Herri, JM: CO_2 capture by hydrate crystallization - a potential solution for gas emission of steelmaking industry. Ener Conv Manag. 48, 1313–1322 (2007).
12. Lin, W, Delahaye, A, Fournaison, L: Phase equilibrium and dissociation enthalpy for semiclathrate hydrate of CO_2 + TBAB. Fluid Phase Equili. 264, 220–227 (2008).
13. Sakamoto, J, Hashimoto, S, Tsuda, T, Sugahara, T, Inoue, Y, Ohgaki, K: Thermodynamic and Raman spectroscopic studies on hydrogen + tetra-*n*-butyl ammonium fluoride semi-clathrate hydrates. Cheml Eng Sci. 63, 5789–5794 (2008).
14. Li, S, Fan, S, Wang, J, Lang, X, Wang, Y: Semiclathrate hydrate phase equilibria for CO_2 in the presence of tetra-*n*-butyl ammonium halide (bromide, chloride, or fluoride). J Chem Eng Data. 55, 3212–3215 (2010).
15. Sun, ZG, Sun, L: Equilibrium conditions of semi-clathrate hydrate dissociation for methane + tetra-*n*-butyl ammonium bromide. J Chem Eng Data. 55, 3538–3541 (2010).
16. Li, DL, Du, JW, Fan, SS, Liang, DQ, Li, XS, Huang, NS: Clathrate dissociation conditions for methane + tetra-*n*-butyl ammonium bromide (TBAB) + water. J Chem Eng Data. 52, 1916–1918 (2007).
17. Ding, Y, Gong, J, Peng, Y: Formation-dissociation characteristics of TBAB hydrate. J China Univ Petrol (Edi of Natu Sci). 35, 150–153 (2011)

18. Mohammadi, AH, Richon, D: Phase equilibria of semi-clathrate hydrates of tetra-*n*-butyl ammonium bromide + hydrogen sulfide and tetra-*n*-butyl ammonium bromide + methane. J Chem Eng Data. 55, 982–984 (2010).
19. Mohammadi, AH, Eslamimanesh, A, Belandria, V, Richon, D: Phase equilibrium of semiclathrate hydrates of CO_2, N_2, CH_4, or H_2 + tetra-*n*-butylammonium bromide aqueous solution. J Chem Eng Data. 56, 3855–3865 (2011).
20. Ye, N, Zhang, P: Equilibrium data and morphology of tetra-*n*-butyl ammonium bromide semiclathrate hydrate with carbon dioxide. J Chem Eng Data. 57, 1557–1562 (2012).
21. Gayet, P, Dicharry, C, Marion, G, Graciaa, A, Lachaise, J, Nesterov, A: Experimental determination of methane hydrate dissociation curve up to 55 MPa by using a small amount of surfactant as hydrate promoter. Chem Eng Sci. 60, 5751–5758 (2005).
22. Makogon, YK: Hydrates of Natural Gas, Tulsa, Oklahoma: PennWell Books (1981)
23. Parent, JS, Bishnoi, PR: Investigations into the nucleation behavior of natural gas hydrates. Chem Eng Commu. 144, 51–64 (1996).
24. Takeya, S, Hori, A, Hondoh, T, Uchida, T: Freezing-memory effect of water on nucleation of CO_2 hydrate crystals. J Phy Chem B. 104, 4164–4168 (2000).
25. Ohmura, R, Ogawa, M, Yasuoka, K, Mori, YH: Statistical study of clathrate-hydrate nucleation in a water/hydrochlorofluorocarbon system: search for the nature of the 'memory effect'. J Phy Chem B. 107, 5289–5293 (2003).
26. Oshima, M, Shimada, W, Hashimoto, TA, Ohgaki, K: Memory effect on semi-clathrate hydrate formation: a case study of tetragonal tetra-*n*-butyl ammonium bromide hydrate. Chem Eng Sci. 65, 5442–5446 (2010).
27. Bishnoi, RJ, Natarajan, V: Formation and decomposition of gas hydrates. Fluid Phase Equili.117, 168–177 (1996).
28. Nixdorf, J, Oellrich, LR: Experimental determination of hydrate equilibrium conditions for pure gases, binary and ternary mixtures and natural gases. Fluid Phase Equili. 139, 325–333 (1997).
29. Masoudi, R, Tohidi, B, Anderson, R, Burgass, RW, Yang, J: Experimental measurement and thermodynamic modeling of

clathrate hydrate equilibria and salt solubility in aqueous ethylene glycol and electrolyte solutions. Fluid Phase Equili. 219, 157–163 (2004).

30. Masoudi, R, Tohidi, B, Danesh, A, Todd, AC: A new approach in modeling phase equilibria and gas solubility in electrolyte solutions and its applications to gas hydrates. Fluid Phase Equili. 215, 163–174 (2004).

31. Sa, JH, Lee, BR, Park, DY, Chun, HC, Lee, KH: Amino acids as natural inhibitors for hydrate formation in CO_2 sequestration. Environ Sci Technol. 45, 5885–5891 (2011).

32. Deschamps, J, Dalmazzone, D: Dissociation enthalpies and phase equilibrium for TBAB semi-clathrate hydrates of N_2, CO_2, $N_2 + CO_2$ and $CH_4 + CO_2$. J Therm Anal Calorim. 98, 113–118 (2009).

33. Oyama, H, Shimada, W, Ebinuma, T, Kamata, Y, Takeya, S, Uchida, T, Nagao, J, Narita, H: Phase diagram, latent heat and specific heat of TBAB semiclathrate hydrate crystals. Fluid Phase Equili. 234, 131–135 (2005).

Chapter 3

Gas Hydrate Stability and Sampling: The Future as Related to the Phase Diagram

E. Dendy Sloan, Carolyn A. Koh, and Amadeu K. Sum

Center for Hydrate Research, Chemical Engineering Department, Colorado School of Mines, 1500 Illinois Street, Golden, CO 80401, USA

ABSTRACT

The phase diagram for methane + water is explained, in relation to hydrate applications, such as in flow assurance and in nature. For natural applications, the phase diagram determines the regions for hydrate formation for two- and three-phase conditions. Impacts are presented for sample preparation and recovery. We discuss an international study for "Round Robin" hydrate sample preparation protocols and testing.

INTRODUCTION

Science in the gas hydrate community has become bifurcated into physical chemistry with applications inside pipelines and in storage, and into geology/geophysics with applications in Nature. While such separation leads to progress with in-depth specialization, relating both communities can provide insight into physical phenomena of measuring and characterizing gas hydrates.

This work shows how the hydrate stability region in the earth and deep ocean is related to the temperature and pressure determined by the thermodynamic phase diagram. We address the question, "What is the stability temperature and pressure region if hydrates are formed from (a) methane-saturated water, (b) free gas and free water, or (c) other phases, such as water-saturated methane gas or liquid?"

Because most of the source gas for hydrates in nature is biogenic [1], characterized by high (>99 mol %) methane content, we consider only the hydrate thermodynamic stability of methane andwater to determine reservoir stability conditions, with implications for sampling. However, the concepts presented may be extended to the thermogenic gas hydrates containing ethane, propane, *etc.*, with the use of a more sophisticated phase prediction program such as Multiflash™, PVTsim®, dbrHydrate®, HYSYS®, or CSMGem [1].

TWO COMMON HYDRATE STABILITY CONDITIONS IN THE EARTH/OCEAN

Consider the most common hydrate stability curves in Figures 1a and b, from the inaugural work of [2]. This diagram represents hydrate stability conditions such as in the 2009 Gulf of Mexico logging-while-drilling expedition [3, 4]. The results of that expedition indicated the three-phase condition determined the hydrate stability region, abetted by sand lithology which allowed free gas accumulation to form hydrates from water. That is, sufficient gas was present to form a separate gas phase.

In both Figure 1a and 1b plots, the non-vertical dashed lines represent thermal gradients with depth, (1a) above and below the permafrost, and (1b) above and below the water-sediment mudline in the ocean. In each figure, solid lines are determined from the three-phase (liquid water + hydrate + methane vapor) thermodynamic stability boundary discussed in the next section.

In Figures 1a and 1b, the upper and lower intersections of thermal gradients with the three-phase boundary determine the depth of feasible hydrate occurrence, shown as shaded yellow regions. In Figure 1a, above the permafrost depth, hydrates exist in equilibrium with ice as the free water phase, although it is difficult to distinguish between ice and hydrate phases. Figure 1b has the hydrate stability region shaded below the mudline (water/sediment line); above the mudline any hydrates will float away because their density is about that of ice, 0.9 g/cc.

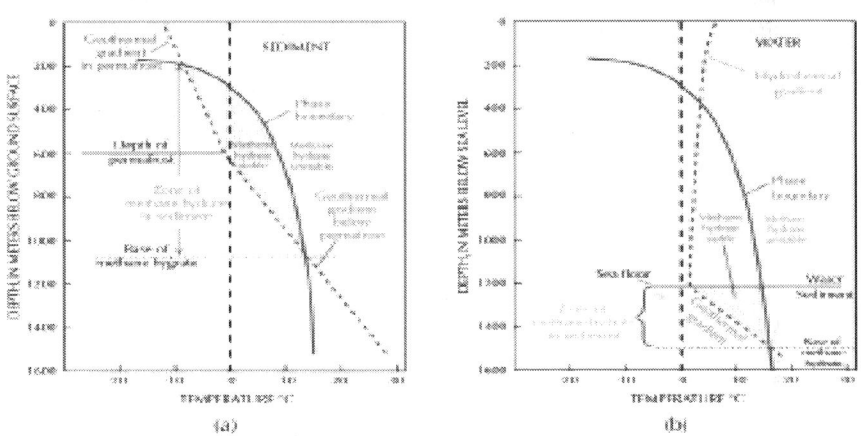

Figure 1: Three-phase (liquid water + hydrate + vapor) stability conditions: (a) in the permafrost and (b) in the ocean [2, 14].

To establish stability depths in Figure 1, we note that there is only one stability temperature at every pressure, or depth, in the phase boundary. This fact was first shown through the general thermodynamic phase analysis of the best-known 19th century American scientist J. Willard Gibbs, for whom the Gibbs Phase Rule is named, as presented in Equation (1):

$$F = C - P + 2 \tag{1}$$

where F = number of intensive variables (e.g., T, P, or phase concentration(s)), C = number of components (e.g., methane and water (or fixed 3.5 wt% salt seawater)), and P = the number of phases (e.g., liquid water (or ice) + hydrates + vapor).

From Equation (1) we conclude that with two components and three phases, the system is univariant, that is, given the temperature (or pressure), the other condition—pressure (or temperature) is fixed, along with the variables of each phase.

Recently, geologists and geophysicists [5,6] have suggested an alternative description for hydrate stability in the ocean with depth. Figure 2 shows a second stability region within the boundary of the three-phase stability shown in Figure 1a. In Figure 2, the lower horizontal line (BGHS) marks the depth of methane hydrate stability in seawater; below this depth free gas is in equilibrium with seawater. The grey region is within the three-phase (liquid water + hydrate + vapor) stability zone, marked by the bottom of the gas hydrate stability zone (GHSZ), with the top marked by the mudline at the seafloor.

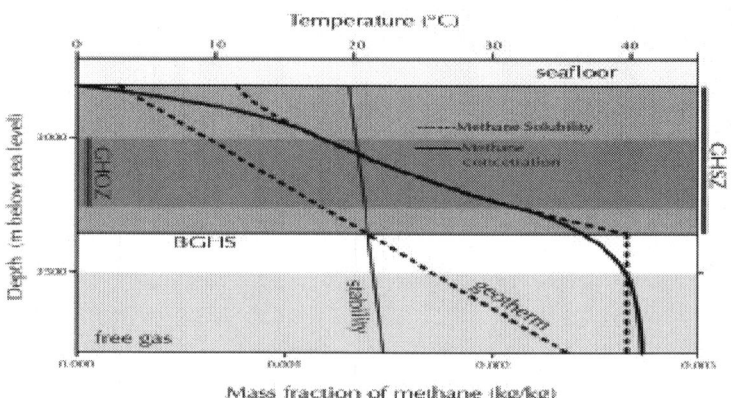

Figure 2: Hydrate Stability Zones in the Ocean.

However, within the GHSZ there is a second, narrower stability region, the darker-shaded gas hydrate occurrence zone (GHOZ), which determines methane solubility coincidence with the three-phase solid boundary, shown in Figure 1b. The narrow depths of the darker GHOZ represent the two-phase (liquid water + hydrate) region in which methane-saturated water is in equilibrium with hydrate.

In the GHOZ a free gas phase is absent, but the methane concentration in water is very small, typically one or two molecules per thousand molecules of water. This situation may occur in nature when the generation rate of methane is too small to provide a separate gas phase, as in the case of anaerobic decomposition of organic matter (producing biogenic methane) at low temperatures of 4 °C.

For the GHOZ two-phase stability, the Gibbs Phase Rule in Equation (1) requires a second variable, other than temperature to be specified, in order to determine the pressure or depth of formation. This second variable is the methane concentration in the sub-ocean porewater. Alternatively, one could specify pressure (or depth) and methane concentration to predict the GHOZ temperature. The two-phase GHOZ region represents the hydrate sample preparation technique of Tohidi et al. [7] and Spangenberg et al. [8] who have prepared hydrate samples from methane-saturated water. Since the methane concentration is so low, this technique requires substantial time for hydrate accumulation.

One purpose of this article is to present a second, thermodynamic explanation of Figures 1 and 2 as an aid in natural hydrate discovery and sampling. As will be shown, the composition of the hydrate differs slightly when formed in either of the two regions of Figure 2. A second purpose of the work is to show the need for adequate hydrates samples in both scenarios, and in hydrate laboratories to ensure reproducible and consistent physical properties.

THE ISOBARIC THERMODYNAMIC PHASE DIAGRAM

Consider the isobaric (constant pressure), temperature *versus* concentration of methane in the equilibrium phases (T-x) in Figure 3 as first published by Huo et al., [9] using spectroscopic and diffraction evidence for the hydrate phase compositions. The diagram is qualitative, so that the single liquid water (L_W) and hydrate (H) phase regions can be observed in visible proportions. In quantitative reality, the L_W region would almost coincide with the leftmost ordinate, due to the small concentration of methane in water; and the H region is so narrow, that it would be quantitatively displayed as a vertical line.

The vapor (V) region at the top of the diagram constitutes the third single phase area of the diagram, in addition to L_W and H. At very low temperatures, e.g., below point 11 in Figure 3, the vapor phase can be considered as a methane-rich liquid phase (L_M). All the other areas in the diagram represent regions of two-phase stability, as marked. The horizontal lines, however, represent temperatures of three-phase stability at the diagram pressure, with the topmost (solid line) at the stability temperature of L_W + H + V, the second highest horizontal line represents I + L_W + H (note that ice is pure), the third horizontal line represents H + V + L_M, and the lowest horizontal line represents H + L_M + M (where M = pure solid methane).

For better comprehension of the phase diagram in Figure 3, it is instructive to do a thought experiment of cooling a vapor mixture of 60:40 mole ratio methane/water from a high temperature at constant pressure, as might be done with the sample in a piston, inside a temperature-controlled bath. The vapor exists as a single-phase until the water dew point (Point 1) is reached, where the composition of the equilibrium liquid water with a little methane dissolved corresponds to Point 5. Further cooling of the gas-liquid mixture causes the amount of the water phase to increase; note that by Gibbs Phase Rule [Equation (1)], two intensive variables (e.g., the P of the entire isobaric diagram, and T) are required to specify the vapor and liquid compositions at the two-phase boundaries.

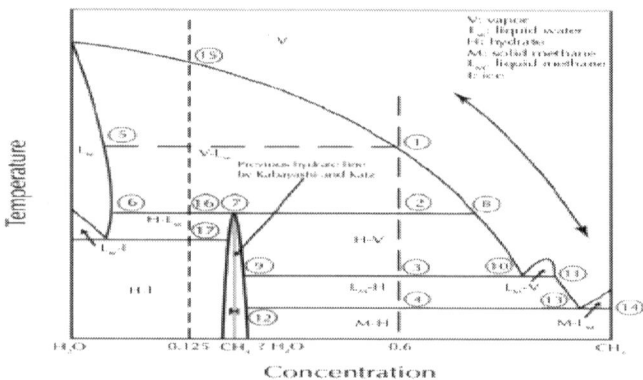

Figure 3: Isobaric (constant pressure) temperature-concentration phase diagram for equilibrium phases present for systems containing methane (M) and water.

In our thought experiment, cooling the system is continued until the temperature of Point 2, where the hydrate phase (vertical area which begins at Point 7) forms in equilibrium with the vapor (Point 8) and liquid (Point 6). At Point 2 three phases (L_W + H + V) coexist for two components, so the Gibbs Phase Rule indicates that only the isobaric pressure of the entire diagram is necessary to specify the unique temperature and concentrations of the three phases (L_W, H, and V) in equilibrium.

At the three-phase condition, typical calculated methane mole fractions in the aqueous, sI hydrate and vapor phases are 0.0014 (Point 6), 0.14 (Point 7), and 0.9997 (Point 8), respectively, showing that the aqueous and vapor regions in Figure 3 are expanded for illustration purposes. Note that the isobaric three-phase temperature at Point 2 marks one P-T condition on the three-phase line (L_W + H + V) shown in Figure 1. In both Figures 1 and 3, at temperatures above this line hydrates cannot form at the specified pressure.

Further heat removal at constant pressure will result in the complete conversion of the free water phase to hydrate at the same initial overall composition of 60 mole% methane. The system enters the two-phase (H + V) region just below the horizontal line at Point 2. By specifying the water composition of the vapor in the two-phase (H + V) region (along the negatively sloping line between Points 8 and 10), one determines how "dry" the gas must be to prevent the possibility of hydrate formation.

At still lower temperatures of the original mixture, some of the vapor condenses to liquid methane at the three-phase (H + V + L_M) boundary (Point 3). Again the three-phase temperature and phase compositions (Points 9, 10, and 11) are specified by the single variable of pressure (F = C − P + 2 = 2 − 3 + 2). Below this three-phase line the vapor phase is totally condensed to a liquidresulting in a two-phase (H + L_M) region between Points 3 and 4, with the phase concentrations given by the borders of the L_M + H region.

Point 4 is the temperature of the lowest three-phase line (H + L_M + M), which occurs just below the solidification point of pure methane (M). Below this line (connecting Points 12, 4, 13, and 14) the liquid methane phase disappears and hydrate exists only with solid methane.

A similar thought experiment line is shown for an overall concentration on the opposite side of the hydrate boundary, by the vertical dashed line at approximately 0.125 methane mole fraction. In this case the dew point (V + L_w) would be realized at Point 15, the three phase (L_w + H + V) line is at Point 16—the unique, above specified phase conditions at Points 6, 7, and 8, and a second three-phase condition (I + L_w + H) is realized at Point 17. Between Points 16 and 17, hydrates exist in equilibrium with methane-saturated (dissolved methane) water. This is the condition shown in Figure 2 for the GHOZ, as well as the two-phase condition at which Tohidi *et al.*, [7] and Spangenberg *et al.*, [8], formed their artificial hydrate samples (though with significant synthesis challenges, as mentioned earlier, and in the later sample preparation section).

HOW DO TWO-PHASE AND THREE-PHASE HYDRATES DIFFER?

Hydrate compositions differ when formed from a single phase, like methane-saturated water, relative to those formed from both gas and liquid water. Subramanian *et al.*, [10] measured both hydrate concentrations in a quartz cell, shown at the top in Figure 4, superimposed above the lower left portion of the phase diagram of Figure 3.

In the Figure 4 top image, a dark hydrate film is shown at the interface of the methane gas and liquid water phases. The rightmost vertical curved arrow connects the hydrate film to the three-phase hydrate concentration on the lower phase diagram. The hydrate film forms rapidly at the interface, at a linear growth rate of about 1 mm per 3 seconds, providing a solid film that prevents contact of the gas with the liquid phase.

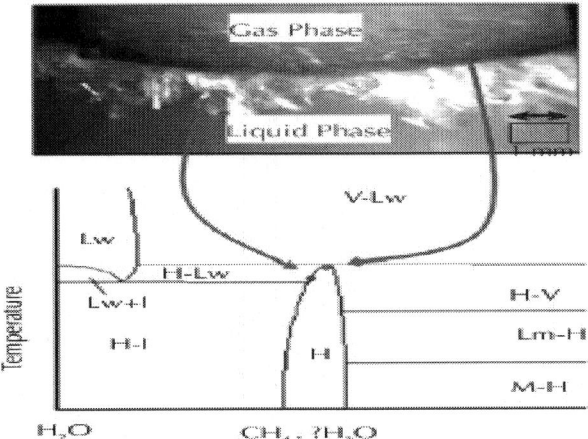

Figure 4: Two-phase and three-phase hydrate formation in a high pressure visual cell (top) connected to the lower left-portion of the phase diagram in Figure 3.

After the formation of the hydrate barrier to further contact of the gas and liquid, excess methane concentration in the liquid forms the dendritic growth shown in the lower portion of the image in Figure 4. The leftmost curved arrow connects the concentration of a dendrite to the methane hydrate concentration on the lower phase diagram. In Figure 4, arrowheads show the position of the two differing hydrate concentrations.

Figure 5 shows the relative methane concentration in the hydrate for both types of hydrate sample in Figure 4, measured by Raman spectroscopy. The top spectrum of Figure 5 is for three-phase hydrates, while the lower spectrum is for two-phase hydrates. The occupancy ratio ($\Theta l/\Theta s$) gives the measured filling of the large cavities to the small cavities by methane in hydrate structure I (sI), obtained by deconvoluting and integrating the large and small peaks of each spectrum, and accounting for the fact that there are three times as many large as small cages in each unit crystal.

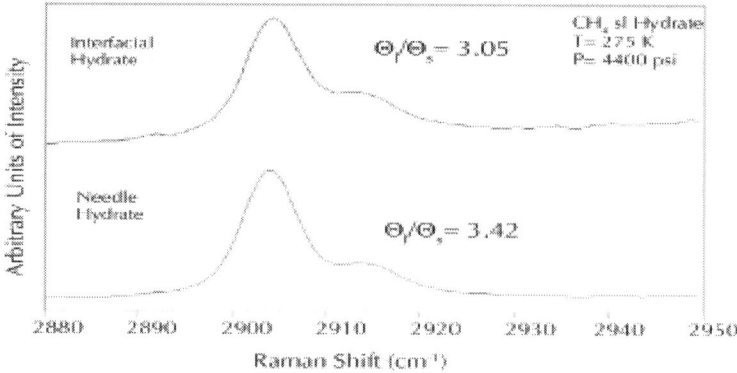

Figure 5: Raman spectra for filling of the large (left peak) and small (right peak) cavities for methane hydrates in three-phase (top spectrum) and two-phase (bottom spectrum) conditions.

The large cavities are typically completely filled [11], so these differences from this ratio indicate the smaller cavities contain 12.8% less methane in the two-phase formation than in three-phase formation. The implication of this result can be far-reaching, especially when considering estimates of methane in natural hydrates. One of the most conservative estimates by Milkov et al., [12] suggest there are 2.5 × 1015 m3 methane at STP in natural hydrates if based on the three-phase hydrates. If all natural hydrates formed in under the two-phase condition (e.g., from methane-saturated water) there may be only 2.2 × 1015 m3 methane at STP in natural hydrates. Such a difference could fuel the USA for 10,000 years at the current national usage rate of 0.68 × 1012 m3 per year. While only a small fraction of the natural methane will be recovered, this example illustrates the effect of concentration differences, and utility of the phase diagram.

The hydrate concentration differences depending on the phases in equilibrium suggest a number of important questions. First, do differences in filling the smaller cavities cause differences in hydrate sample properties other than concentration? How common are the natural hydrate environments which do not generate sufficient methane for a separate gas phase? These questions the hydrate community has yet to address. The reader is referred to the excellent review article on hydrate properties by Waite et al., [13] for the existing state of the art in natural hydrates properties.

SAMPLE PREPARATION AND ROUND-ROBIN TESTING

During the last 20 years, the hydrate community has attempted to make samples of hydrate in sediments, with only moderate success. Three- and two-phase hydrate formation methods have been used to synthesize hydrate-bearing sediment (HBS) samples in attempts to replicate naturally occurring HBS samples. Three-phase formation methods (liquid water + hydrate + free gas) are most widely applied, because they are technically simpler and far less time-consuming than two-phase formation (dissolved gas in water + hydrate). Three-phase formation methods include free gas being added to partially water-saturated sediment, or to ice grains + sediment. Two phase formation methods use dissolved gas in water + sediment [14,15]. There remains a dichotomy in that three-phase formation methods can readily produce HBS samples with high hydrate saturations, which is challenging to achieve using two-phase formation; but three-phase formation tends to produce samples which are less representative (in terms of pore-scale structure and physical properties) of natural HBS compared to samples prepared using two-phase formation.

Two questions must be addressed for these samples:
- How reproducible are artificial samples in sediments, both inter-laboratory and intra-laboratory? For samples without sediments, the preparation standard is the method of Stern et al., [15]. This technique converts small ice particles to hydrates under high methane pressures, using temperature traverses of the ice liquidus; pressure change is monitored as an indication of hydrate formation completion.
- How representative of natural hydrate samples are artificial hydrate samples? Problems of hydrate homogeneity arise in both instances, due to hydrate formation at the interface, which blocks further phase contact of the vapor-liquid (three-phase formation).

The authors (in collaboration with W. Waite, USGS and J.C. Santamarina, Georgia Tech) have proposed an initial attempt to address the first question [16] using multi-laboratory, round-robin sample preparations and measurements. The objective is to establish standards and methodologies to synthesize reproducible inter-laboratory hydrate samples. We propose that many laboratories perform and compare the results of five tests using standard sediment (F110 Ottawa sand provided by USGS). All tests will be based on measuring compressional (V_p) and shear wave speed (V_s) in unconsolidated sand with three primary variables: (1) porosity (between ~36% and 40% ± 1%), (2) effective stress (K_o = horizontal stress/vertical stress ~0.45), and (3) pore-space saturation (20% at 5 °C).

Inter-laboratory results will compare five tests, at conditions agreed upon at the 2010 Hydrate Gordon Conference in Maine, and Fiery Ice Workshops in Wellington:

- Dry F110 Ottawa Sand with two porosities: Porosity 1 will be 36% and porosity 2 will be 40% (±1%). The effective vertical and horizontal stresses will be: 3 MPa and 1.35 MPa, respectively, at a temperature of 5 °C, with an initial pore space water saturation of 0%.
- Water-saturated F110 Ottawa Sand: Porosity 1 will be 36% and porosity 2 will be 40% (±1%). The effective vertical and horizontal stresses will be: 3 MPa and 1.35 MPa, respectively, at a temperature of 5 °C, with an initial pore space water saturation of 100%.
- Partially water-saturated F110 Ottawa Sand: Porosity 1 will be 36% and porosity 2 will be 40% (±1%). The effective vertical and horizontal stresses will be: 3 MPa and 1.35 MPa, respectively, at a temperature of 5 °C, with an initial pore space water saturation of 20%.
- Frozen partially water-saturated F110 Ottawa Sand: Porosity 1 will be 36% and porosity 2 will be 40% (±1%). The effective vertical and horizontal stresses will be: 3 MPa and 1.35 MPa, respectively, at a temperature of -5 °C, with an initial pore space water saturation of 20%.
- Partially hydrate-saturated F110 Ottawa Sand: Porosity 1 will be 36% and porosity 2 will be 40% (±1%). The effective vertical and horizontal stresses will be: 3 MPa and 1.35 MPa, respectively, at a temperature of 5 °C, with an initial pore space water saturation of 20%. Each laboratory will specify their method of introducing gas, and will provide imagery.

The results from the above Tests 1-5 will be presented at the Seventh International Conference on Gas Hydrates in Edinburgh, U.K., July 17–

21, 2011. Future tests will be on end members, where results (V_s, V_p) are compared with natural hydrate samples/well-logs, *etc.*, such as (a) cementing/strengthening (of sediment) hydrates, (b) pore-filling/load-bearing hydrates, (c) with hydrates as part of fluids, (d) hydrate layers/nodules/veins, (e) hydrate layers formed from auto-layering of broader grain size distribution of F110 Ottawa sand, (f) checking morphology using micro-CT X-ray imaging, cryo-SEM.

CONCLUSIONS

The goal of this article is twofold: (1) To show how hydrate samples are affected by thermodynamics, as specified by the phase diagram, and (2) to state the need for a uniform hydrate sample-in-sediment protocol, and the initial attempts to address that need via a round-robin testing among a number of laboratories. It has been shown that temperatures, pressures, and phase compositions can be determined by the number of phases present when hydrates are formed. Yet to be answered are questions regarding how changes in hydrate composition affect hydrate properties, and if it is possible to provide artificial hydrate samples which are reproducible and representative of hydrates in nature.

REFERENCES

1. Sloan, E.D.; Jr.; Koh, C.A. *Clathrate Hydrates of Natural Gases*, 3rd ed.; CRC Press: Boca Raton, FL, USA, 2008.
2. Kvenvolden, K.A. Methane Hydrate—A Major Reservoir of Carbon in the Shallow Geosphere? *Chem. Geol.* 1988, *71*, 41–51.
3. Collett, T.; Boswell, R.; Frye, M.; Shedd, W.; Godfriaux, P.; Dufrene, R.; McConnell, D.; Mrozewski, S.; Guerin, G.; Cook, A.; Jones, E.; Rana, R. Gulf of Mexico Gas Hydrate Joint Industry Project Leg II: Logging-While-Drilling Operations and Challenges. In *Proceedings of the Offshore Technology Conference*, Houston, TX, USA, May 2010. 4.
4. Boswell, R. Is Gas Hydrate Energy Within Reach? *Science* 2009, *325*, 957–958.

5. Tréhu, A.M.; Glemings, P.B.; Bangs, N.L.; Chevallier, J.; Gracia, E.; Johnson, J.E.; Liu, C.-S.; Liu, X.; Riedel, M.; Torres, M.E. Feeding methane vents and gas hydrate deposits at south Hydrate Ridge. *Geophys. Res. Lett.* 2004, *31*, L23310.

6. Tréhu, A.M.; Ruppel, C.; Holland, M.; Dickens, G.R.; Torres, M.E.; Collett, T.S.; Goldberg, D.S.; Riedel, M.; Schultheiss, P. Gas hydrates in marine sediments: lessons from scientific ocean drilling. *Oceanography* 2006, *19*, 124–142.

7. Tohidi, B.; Anderson, R.; Clennell, B.; Yang, J.; Bashir, A.; Burgass, R. Application of High Pressure Glass Micromodels to Gas Hydrates Studies. In *Proceedings of the Fourth International Conference on Gas Hydrates*, Yokohama, Japan, May 2002.

8. Spangenberg, E.; Beeskow-Strauch, B.; Luzi, M.; Naumann, R.; Schicks, J.M.; Rydzy, M. The Process of Hydrate Formation in Clastic Sediments and Its Impact on Their Physical Properties. In *Proceedings of the Sixth International Conference on Gas Hydrates*, Vancouver, BC, Canada, July 2008.

9. Huo, Z.; Hester, K.; Miller, K.T.; Sloan, E.D., Jr. Methane Hydrate Non-Stoichiometry and Phase Diagram. *AIChE J.* 2003, *49*, 1300–1306.

10. Subramanian, S.; Sloan, E.D., Jr. Microscopic Measurements and Modeling of Hydrate Formation Kinetics. *Ann. N. Y. Acad. Sci.* 2000, *912*, 583–592.

11. Udachin, K.A.; Ratcliffe, C.I.; Ripmeester, J.A. Single Crystal Diffraction Studies of Structure I, II, H Hydrates: Structure, Cage Occupancy and Composition. *J. Supramol. Chem.* 2003, *2*, 405–408.

12. Milkov, A.V. Global Estimates of Hydrate-Bound Gas in Marine Sediments: How Much is Really Out There? *Earth Sci. Rev.* 2004, *66*, 183–197.

13. Waite, W.F.; Santamarina, J.C.; Cortes, D.D.; Dugan, B.; Espinoza, B.N.; Germaine, J.; Jang, J.; Jung, J.W.; Kneafsey, T.J.; Shin, H.; Soga, K.; Winters, W.J.; Yun, T.S. Physical properties of hydrate bearing sediments. *Rev. Geophys.* 2009, *47*, RG4003.

14. Paul, C.; Reeburgh, W.S.; Dallimore, S.R.; Enciso, G.; Green, S.; Koh, C.A.; Kvenvolden, K.A.; Mankin, C.; Riedel, M. Realizing the Energy Potential of Methane Hydrate for the United States.

National Academies NRC Report; National Academies Press: Washington, DC, USA, 2010.

15. Stern, L.; Kirby, S.; Durhan, W. Peculiarities of Methane Calthrate Hydrate Formation and Solid State Deformation, Including Possible Superheating of Water Ice. *Science* 1996, *273*, 1843–1848.

16. Waite, W.F.; Santamarina, J.C.; Koh, C.A.; Sloan, E.D., Jr.; Grozic, J.L.H.; Hester, K.C.; Howard, J.; Mahajan, D.; Priest, J.; Rydzy, M.; Seol, Y.; Winters, W.J.; Yun, T.S.; Inter-laboratory comparison of wave velocity measurements. In *Proceedings of the Seventh International Conference on Gas Hydrates*, Edinburgh, UK, July 2011.

Chapter 4

Molecular Storage of Ozone in a Clathrate Hydrate: An Attempt at Preserving Ozone at High Concentrations

Takahiro Nakajima[1], Taisuke Kudo[1], Ryo Ohmura[1], Satoshi Takeya[2], and Yasuhiko H. Mori[1]

[1]Department of Mechanical Engineering, Keio University, Yokohama, Japan,
[2]Research Institute of Instrumentation Frontier, National Institute of Advanced Industrial
Science and Technology (AIST), Tsukuba, Japan

ABSTRACT

This paper reports an experimental study of the formation of a mixed $O_3 + O_2 + CO_2$ hydrate and its frozen storage under atmospheric pressure, which aimed to establish a hydrate-based technology for preserving ozone (O_3), a chemically unstable substance, for various industrial, medical and consumer uses. By improving the experimental technique that we recently devised for forming an $O_3 + O_2 + CO_2$ hydrate, we succeeded in significantly increasing the fraction of ozone contained in the hydrate. For a hydrate formed at a system pressure of 3.0 MPa, the mass fraction of ozone was initially about 0.9%; and even after a 20-day storage at −25°C and atmospheric pressure, it was still about 0.6%. These results support the prospect of establishing an economical, safe, and easy-to-handle ozone-preservation technology of practical use.

INTRODUCTION

Ozone (O_3) is known as a powerful oxidant and, due to this nature, it is widely used for, for example, the decontamination of air and water, the sterilization of perishables, the disinfection of medical instruments, and the cleaning or surface-conditioning processes in the semiconductor industry. However, it is neither very easy nor economical to use ozone in consumer applications such as sanitizing foods and drinking water, removing pesticide residues from fruits and vegetables, treating water in aquariums for suppressing bacteria growth, etc. This is because, to artificially generate ozone, we need a high-voltage electric device such as a corona discharger or a cold plasma generator and, once generated, ozone in the gaseous state rapidly decomposes to oxygen (O_2). Thus, it is generally believed that ozone can neither be stored nor transported and must be produced on site. For the limited consumer use of ozone, ozonated water (liquid water in which ozone is physically dissolved) and ozonated ice (water ice holding microbubbles of an ozone-containing gas) are commercially available. However, the ozone concentration in such ozonated water or ice is generally on the order of 1 or 10 ppm even in its fresh state, and rapidly decays with time. The above state of affairs seriously restricts the situations allowing the use of ozone. If we find a convenient means for transporting ozone from its production site to any place where ozone is needed, the utility of ozone will be significantly expanded.

Clathrate hydrates (abbreviated hydrates) are crystalline solid compounds each composed of host water molecules hydrogen-bonded into a structure of interlinked cages. Unless the given pressure is extremely high (typically on the order of gigapascals), each cage contains at most one guest molecule of a substance other than water [1]. That is, the guest molecules in a hydrate are isolated by the cage walls due to van der Waals forces and thereby prevented, in general, from mutual interactions. This indicates that hydrates have a high potential of storing chemically unstable substances, such as ozone, in the form of encaged guest molecules.

The idea of storing ozone in a hydrate was first presented by McTurk and Waller [2], [3] in 1964. They reported the formation of an ozone-containing hydrate in an experimental system containing pure ozone, carbon tetrachloride (CCl_4) and water, and, based on their X-ray diffraction measurement, indicated that this hydrate was a double O_3+ CCl_4 hydrate in structure II. However, they provided neither any quantitative evaluation of the ozone content nor any experimental evidence for actual ozone preservation in their hydrate.

The pure ozone used by McTurk and Waller [2], [3] is not easily available. Besides, it is explosive and very difficult to handle in practice. Carbon tetrachloride is effective as a *help guest* for lowering the pressure required for hydrate formation, though it is toxic and may be unsuitable for some applications. An attempt at forming a hydrate from a dilute ozone-containing gas (a mixture of ~5% O_3 and ~95% O_2 generated from a commercial ozone generator) in the absence of any help guest was reported by Masaoka et al. [4]. They formed a hydrate at a pressure of 13 MPa and a temperature of −25°C, and determined the ozone content of the hydrate to be 2.3 g/L (≈ 0.2% in mass fraction). They also performed a storage test of the hydrate at the same pressure–temperature conditions as those in the formation process, i.e., 13 MPa and −25°C, and observed only a slight decrease in the ozone content of the hydrate during 10-days storage after its formation.

More recently, Muromachi et al. [5] formed a hydrate from a ozone + oxygen gas mixture (~8% in mole fraction of ozone) and carbon tetrachloride or xenon (Xe) at a pressure of 0.25 or 0.35 MPa and a temperature of 0.1°C. They showed that, if cooled to −20°C under an aerated atmospheric-pressure condition, the O_3+ O_2+ CCl_4 and O_3+ O_2+ Xe hydrates could preserve ozone for more than 20 days at mass fractions around 0.2% and 0.1%, respectively. Subsequently, Muromachi et al. [6] performed phase-equilibrium measurements for the O_3+ O_2+ CCl_4 and O_3+ O_2+ CH_3CCl_2F hydrates.

For practical applications of ozone-containing hydrates, the use of a toxic or very expensive substance as the help guest should be avoided. On the other hand, the hydrates can be desirably formed at moderate pressures and preserved at a moderately cooled atmospheric-pressure condition. In order to satisfy these requirements, Nakajima et al. [7] selected carbon dioxide (CO_2) as the help guest, and formed an O_3+ O_2+ CO_2 hydrate from an O_3+ O_2 gas mixture (10−12% in O_3 mole fraction) blended with pure CO_2 in a molar ratio of 1:7 at a pressure of 1.9 MPa and a temperature of 0.1°C. They performed preservation tests with the O_3+ O_2+ CO_2 hydrate at different storage temperatures from −5°C to −30°C under an aerated atmospheric-pressure condition. The results of these tests showed that, for ozone preservation over 20 days at a mass fraction around 0.1%, the storage temperature should be −25°C or lower. This inferiority in ozone-preserving function of the O_3+ O_2+ CO_2 hydrate in comparison with the O_3+ O_2+ CCl_4 hydrate [5] is, as demonstrated by a relevant phase-equilibrium study [8], essentially due to the higher phase-equilibrium pressure for the former than the latter at any given temperature and hence inevitable. Despite such a thermodynamic disadvantage of the O_3+ O_2+ CO_2 hydrate as compared to the O_3+ O_2+ CCl_4 hydrate, the former represents a good compromise between the ozone preservability versus the biological safety and the economy in system operation, and is possibly the best selection as the ozone storage medium for practical use. Moreover, we can expect that the gas mixture released from an O_3+ O_2+ CO_2 hydrate will have some synergetic effect of the ozone and carbon dioxide for sterilizing foods [9] and will possibly be more suitable for food-industrial applications than the O_3+ O_2 or O_3+ air gas mixtures directly generated from commercial ozone generators.

This study is an extension of the first O_3+ O_2+ CO_2 hydrate study by Nakajima et al. [7] discussed above. The major objectives of this study were (a) to generate O_3+ O_2+ CO_2 hydrates having higher ozone fractions, and (b) to perform ozone preservation tests with these hydrates in order to examine their practical utility. The study was successful regarding both of these objectives. We confirmed that, with simple modifications of the hydrate-forming procedure, the initial ozone content of a hydrate can be multiplied severalfold as compared to that previously observed [7], and that more than half of such a high ozone content remains after a 20-day hydrate storage under an aerated atmospheric-pressure condition at a temperature of −25°C.

EXPERIMENTAL SECTION

The general experimental scheme used in this study was the same as that used in our previous study [7] that first dealt with an $O_3 + O_2 + CO_2$ hydrate. However, some core portions of the hydrate-forming apparatus were modified this time in order to increase the water-to-hydrate conversation ratio (i.e., to reduce the fraction of ice in the hydrate + ice solid mixtures for use in ozone preservation tests) and to allow pressurizing the $O_3 + O_2$ gas mixture released from an ozone generator before mixing it with CO_2 gas. Details of the materials, equipment and procedure used in this study are described below.

Materials

The raw materials used for forming the $O_3 + O_2 + CO_2$ hydrates were oxygen certified to the purity of 99.9% (volume basis) and carbon dioxide certified to the purity of 99.995% (volume basis) by their supplier (Japan Fine Products Corp., Kawasaki, Kanagawa Prefecture, Japan), and water deionized and distilled in our laboratory. Oxygen was used for generating an $O_3 + O_2$ gas mixture (>11% in mole fraction of O_3) with the aid of a dielectric-barrier-discharge-based ozone generator (ED-OGS-HP1, EcoDesign Co., Ltd., Saitama Prefecture, Japan).

Apparatus

The experimental setup used to form the hydrates is schematically illustrated in Fig. 1. By comparing this figure to Fig S1 in our previous paper [7], one may realize how the setup used this time had been modified from its predecessor. The hydrate-forming reactor [indicated as (I) in Fig. 1] was completely renewed. It was a pan-type stainless-steel vessel with a 65-mm inside diameter and 100-cm^3 inside volume. An impeller-blade stirrer magnetically connected to the drive shaft of an external, variable-speed, ac motor was inserted into this reactor in order to provide its contents with stronger mixing than a magnetic stirrer inside a tall cylindrical reactor used in the previous setup [7] did. As before, the reactor was immersed in a temperature-controlled bath containing an aqueous ethylene-glycol solution.

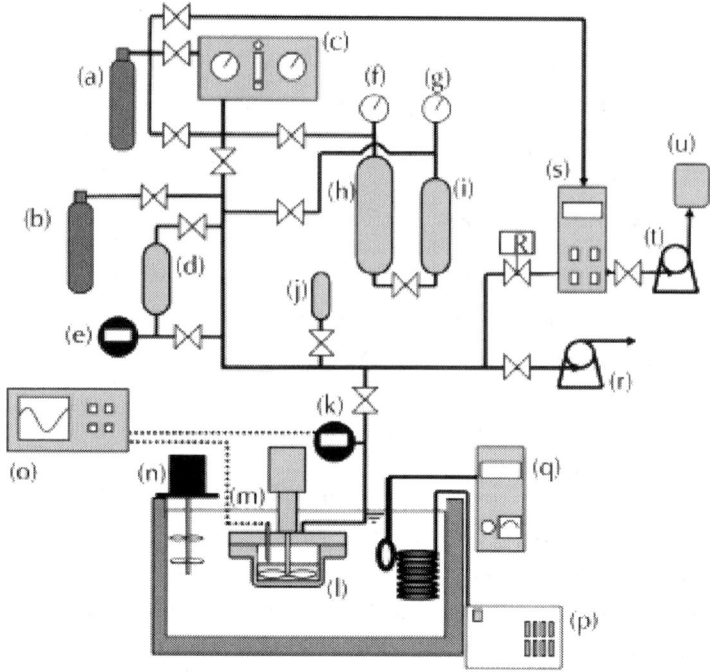

Figure 1: Schematic illustration of the experimental setup for forming the O_3+ O_2+ CO_2 hydrates. This setup consists of (a) an oxygen cylinder, (b) a carbon-dioxide cylinder, (c) an ozone generator, (d) a gas-mixing chamber, (e) a pressure gauge, (f) and (g) pressure gauges, (h) and (i) gas-pressurizing chambers, (j) a gas-sampling chamber, (k) a pressure gauge, (l) a hydrate-forming reactor, (m) a Pt-wire resistance thermometer, (n) a stirrer, (o) a data logger, (p) an immersion cooler, (q) a PID-controlled heater, (r) a vacuum pump, (s) an ozone monitor, (t) a vacuum pump, and (u) an ozone decomposer.

Another modification in the experimental setup was the installation of two gas-pressurizing chambers, which are indicated as (h) and (i) in Fig. 1. They were stainless-steel cylinders with capacities of 3785 cm³ and 2250 cm³, respectively. The smaller chamber (i) was used to store the O_3+ O_2 gas mixture supplied from the ozone generator (c) at a pressure up to about 0.3 MPa, while the larger one (h) was initially charged with water. By injecting oxygen gas supplied from the external high-pressure cylinder (a) into the larger chamber, water could be displaced from the larger chamber to the smaller chamber, thereby increasing the pressure of the O_3+ O_2 gas mixture to the prescribed level.

Procedure

The procedure of forming an $O_3 + O_2 + CO_2$ hydrate using the renewed setup (Fig. 1) is as described below. First, the reactor (l) was charged with ~30 g of water and immersed in a bath of an aqueous ethylene-glycol solution temperature-controlled at 0.1°C. The reactor, the gas-pressurizing chambers (h) and (i), and the gas-mixing chamber (d) were then flushed at least five times with pure oxygen gas, then evacuated. After confirming that the mole fraction of ozone in the gas mixture released from the ozone generator (c) was in the range of 10−12%, the smaller gas-pressurizing chamber (i) and the gas-mixing chamber (d) were charged with this mixture up to a pressure of 0.3 MPa. Oxygen gas from the high-pressure cylinder (a) was then injected into the larger gas-pressurizing chamber (h) to make the water stored in it flow into the smaller chamber (i) and thereby to make the $O_3 + O_2$ gas mixture flow out of the latter chamber into the gas-mixing chamber (d), until the pressure inside the gas-mixing chamber increased to the prescribed level. When the pressure inside the gas-mixing chamber did not sufficiently increase at this stage, the above serial operations beginning with the charging of the smaller gas-pressurizing chamber with a fresh $O_3 + O_2$ gas mixture was repeated until the pressure was raised to the prescribed level. Subsequently, CO_2 gas was supplied to the gas-mixing chamber until the pressure inside increased to 3 or 4 MPa. The $O_3 + O_2 + CO_2$ gas mixture thus prepared was supplied to the hydrate-forming reactor (l) until the pressure inside the reactor increased to the prescribed level (2.0, 2.5 or 3.0 MPa). A series of intermittent batch operations for forming a hydrate was then started by turning on the stirrer in the reactor. For preventing the system pressure from significantly decreasing from the prescribed level and for minimizing the change in composition of the gas mixture inside the reactor, each batch operation was not allowed to continue for a long period but interrupted by a gas-exchange operation for replacing the residual gas mixture inside the reactor by a fresh gas mixture newly prepared in, and supplied from, the gas-mixing chamber (d). Such a change in the batch and gas-exchange operations was repeated several times until no decrease in system pressure was detected during each batch operation and hence we judged that the hydrate formation had already ceased. The reactor was then cooled to −15°C to freeze the contents of the reactor except for the residual gas mixture. The gas mixture was then discharged into a 50-cm³ gas-sampling chamber (j) for its compositional analysis using a gas chromatograph (Agilent 3000 Micro Gas Chromatograph). After removing the reactor from the bath and dipping it into a liquid-nitrogen pool, the formed hydrate was removed from the reactor and crushed into particles with a 5−7 mm linear dimension. A small portion (~1−2 g in mass) of the hydrate was sampled for an iodometric measurement for determining its initial ozone content. The rest of the hydrate was stored in a Pyrex test tube, if it was to be used for a subsequent preservation test.

The procedure of the ozone preservation tests performed in this study was completely the same as that employed in our previous studies [5. 7], hence its description is omitted here. The technique used in the PXRD measurement with some hydrate sample was also the same as that described elsewhere [7].

RESULTS AND DISCUSSION

Before presenting the results of a series of ozone preservation tests performed in this study, we need to specify the actual contents of what we have called "hydrates" and "ozone fractions" in the preceding sections in relation to the previous hydrate-preservation studies [4], [5], [7]. Because liquid water used for forming hydrates in any of these studies could not be completely converted to a hydrate, each preservation-test specimen prepared by cooling the formed hydrate, together with residual water, to a test temperature below 0°C must have been a mixture of the hydrate and water ice. Inevitably, the "ozone faction" measured by some macroscopic means (e.g., the iodometric technique [5], [7]) is not an intrinsic ozone fraction of the hydrate but an *effective* fraction defined as the ratio of the mass of ozone contained in a given hydrate + ice mixture to that of the mixture itself. This means that an increase in such an effective ozone fraction, x_{O3}, may be achieved either by (i) decreasing the ice fraction in the hydrate + ice mixtures for storing ozone or by (ii) increasing the true ozone fraction in the hydrate. As described in the Experimental Section, we attempted at realizing both of these means in order to significantly increase x_{O3} as compared to its magnitude (~0.1%) observed in the previous study of this series [7]. The former means was realized by intensifying the stirring of the gas/liquid contents in the hydrate-forming rector, while the latter was carried out by varying the system pressure as well as the composition of the feed gas supplied to the reactor during each hydrate-forming operation. Varying the feed-gas composition was made by controlling the ratio of CO_2 addition to the $O_3 + O_2$ mixture generated from an ozone generator at a nearly fixed O_3 fraction (10–12%).

Figure 2 shows a powder X-ray diffraction (PXRD) pattern (measured at 98 K) of an $O_3 + O_2 + CO_2$ hydrate formed from a mixture of $O_3 + O_2$ and CO_2 in a nearly 2:8 molar ratio at a system pressure p of 2.0 MPa and a temperature T of 0.1°C. This pattern indicates that the hydrate sample used here was a mixture of a hydrate in structure I (sI) with the lattice constant of 11.8294(4) Å and water ice in two different crystal forms, i.e., hexagonal ice, Ih, and cubic ice, Ic. The mass fraction of the hydrate was estimated to be 0.89, which was significantly higher than the corresponding estimate (~0.3) for the $O_3 + O_2 + CO_2$ hydrate samples formed in the previous study [7].

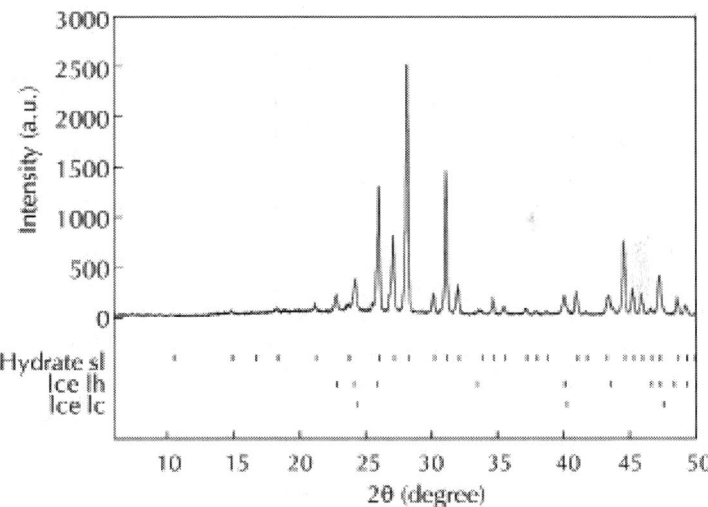

Figure 2: PXRD profile of an $O_3 + O2 + CO_2$ hydrate at 98 K the solid curve shows the intensities observed using Cu−Kα radiation. The top row of tick marks represent the calculated peak positions for the structure I hydrate, and the lower two rows represent those for the hexagonal ice Ih and cubic ice Ic, respectively. The hydrate sample (accompanied by ice crystals) used in this PXRD measurement was formed from a mixture of $O_3 + O2$ and CO_2 in a nearly 2: 8 molar ratio at the condition of p = 2.0 MPa and T = 0.1°C.

We performed hydrate-forming experiments at three different system pressures (2.0, 2.5 and 3.0 MPa) and four different $O_3 + O_2$ versus CO_2 molar ratios (1: 9, 2: 8, 3:7 and 4: 6, each accompanied by slight run-to-run scatter) in the feed gas. The hydrate formed in each experiment was subjected to an iodometric measurement to determine its ozone content, i.e., the initial X_{o3} value for the hydrate which we denote $X_{o3/inte}$ hereafter. Figures 3 and 4 show the variations in $X_{o3/inte}$ depending on the system pressure p for each hydrate-forming operation and the feed-gas composition, respectively, in which the feed-gas composition is represented by X_{o3}, the mole fraction of ozone in the gas phase in contact with the hydrate at the end of each hydrate-forming operation (consult Tables S1 and S2 and Fig. S1 in Supporting Information S1 for $x_{O3'init}$ the complete sets of $X_{o3/inte}$ and X_{o3} data and the graphical plots of the $X_{o3/inte}$ data) We note that $X_{o3/inte}$ shows no systematic dependence on the system pressure (Fig. 3) but a quasi-linear dependence on X_{o3} (Fig. 4). This fact indicates that $X_{o3/inte}$ was primarily controlled by the competitive fractional filling of the hydrate cages by O_3, O_2 and CO_2 molecules and that most of the hydrate cages were occupied by some of these guest molecules even at the lowest system pressure, $p = 2.0$ MPa, prescribed in the present experiments. Our estimation of the cage occupancies by O_3, O_2 and CO_2 molecules is described in Supporting Information S2. Consult Table S3 and Fig. S2 in Supporting Information S2 for the estimated occupancy values

Molecular Storage of Ozone in a Clathrate Hydrate: An Attempt...

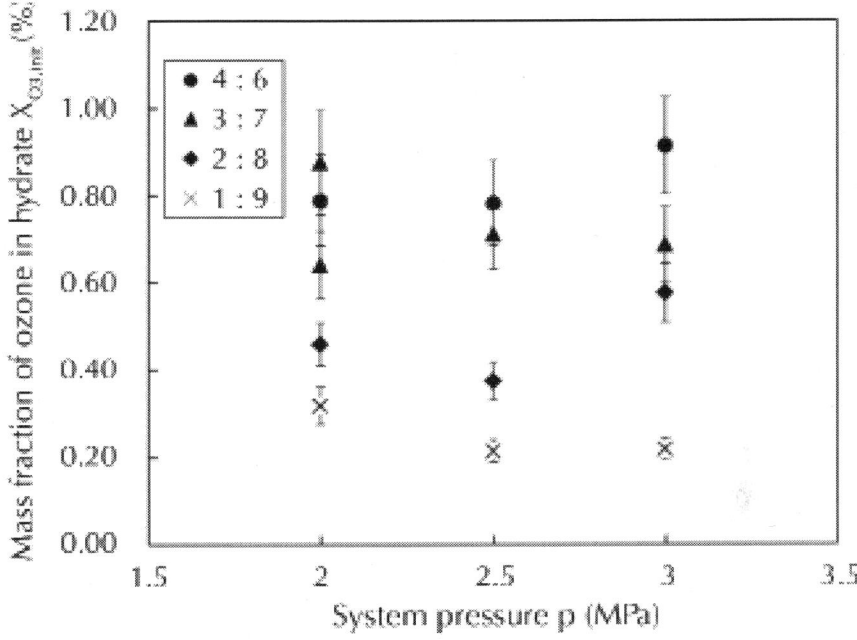

Figure 3: The initial ozone fraction in the formed hydrate versus the system pressure. The legend inserted in the graph indicates the $O_3 + O_2$ versus CO_2 molar ratio in the feed gas used for each operation. Each data point represents the arithmetic mean of the three $x_{O3,init}$ values obtained for the different hydrate samples. The error bar for each data point represents the uncertainty of the ozone-fraction measurement by iodometry.

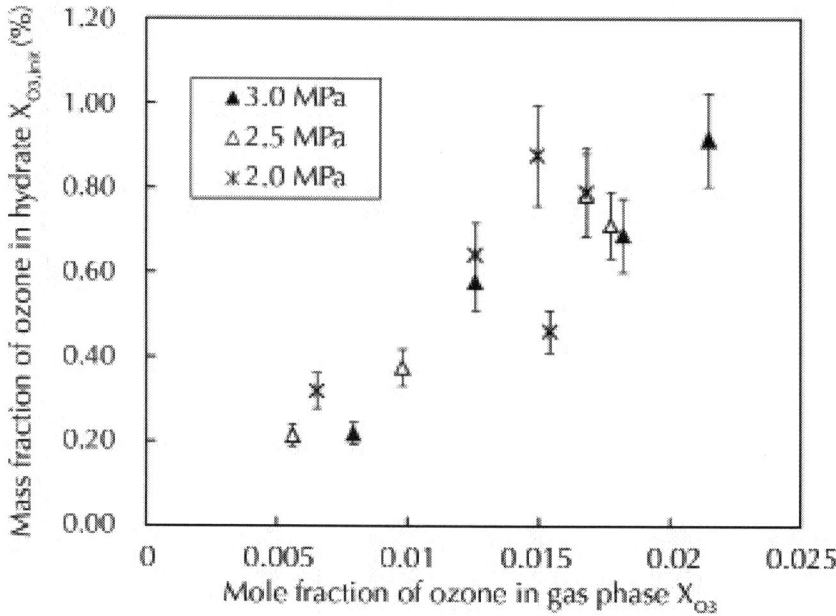

Figure 4: The initial ozone fraction in the formed hydrate versus the gas-phase composition. The mole fraction of ozone, X_{O3}, shown here is for the gas phase inside the reactor when the hydrate formation ceased. The legend inserted in the graph indicates the system pressure p during each hydrate-forming operation. The error bar for each data point represents the uncertainty of the ozone-fraction measurement by iodometry.

For the hydrates formed at the two higher system pressures, p = 2.5 and 3.0 MPa, ozone preservation tests were performed following the procedure employed in the previous relevant studies [5], [7]. The hydrates were stored in an aerated atmospheric-pressure (0.101 MPa) condition temperature-controlled at −25°C. Figure 5 shows the results of these tests, i.e., two x_{O3}-data sets each obtained by continually sampling the stored hydrate for iodometric measurements during a period extending to 20−26 days after the formation of the hydrate. In addition, Fig. 5 shows, for comparison, two x_{O3}-data sets previously obtained with hydrates formed at a lower system pressure and a lower (O_3+ O_2)-to-CO_2 ratio. Obviously, the data obtained in the present preservation tests show much higher x_{O3} values than the previous data through the entire hydrate-storage period. The former exhibited x_{O3} values of 0.4−0.6% at stages of ~20-days storage, which is a few times higher than the values exhibited by the latter at the same stages. The relatively sharp decrease in x_{O3} observed in the present tests, particularly during the initial several days of storage, is presumably ascribable to the higher $O_3 + O_2$ fractions (or the lower CO_2 fractions) in the hydrates used in the tests as compared to the hydrates used in the previous study [7]. Because the thermodynamic equilibrium condition shifts toward a higher pressure or lower temperature with an increase in the $O_3 + O_2$ fraction, or a decrease in the CO_2 fraction, in the hydrate [8], we can reasonably assume that the hydrates used in the present preservation tests suffered higher thermodynamic driving forces for their dissociation than those used in the previous study [7] under the same storage condition. In addition, the lower ice fraction mixed with the hydrates used in the present tests possibly weakened the ice-barrier effect for suppressing the hydrate dissociation. As recognized in Fig. 5, the present test data show an asymptotic decrease in x_{O3} with time, with an apparent half-life period of 20−25 days. This period is considered to be long enough to allow the practical use of ozone-containing hydrates for most industrial, medical and consumer needs.

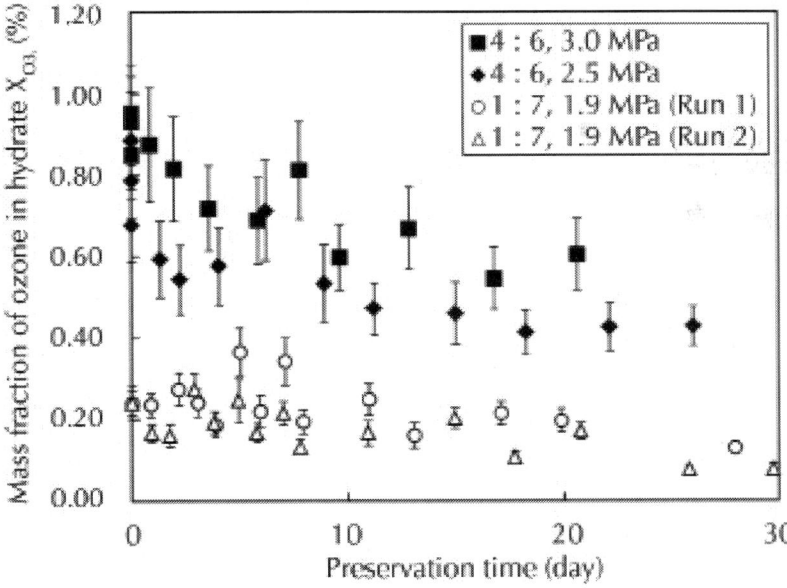

Figure 5: Results of the ozone preservation tests. This graph shows the time evolution of ozone fraction (mass basis) in each $O_3 + O_2 + CO_2$ hydrate stored

under an aerated atmospheric-pressure (0.101 MPa) condition temperature-controlled at −25°C. The comparison of the ozone preservation test data obtained in this study (marked by closed symbols) and those from a previous study [7] (marked by open symbols) are compared. The legend inserted in the graph indicates the $O_3 + O_2$ versus CO_2 molar ratio in the feed gas and the system pressure p for each hydrate-forming operation. The error bar for each data point represents the uncertainty of the ozone-fraction measurement by iodometry.

CONCLUSIONS

This study demonstrated that ozone can be stored up to a mass fraction of ~0.9% in a structure-I hydrate (containing water ice by ~10%) formed from a ternary (ozone + oxygen + carbon dioxide) gas mixture containing ozone up to a mole fraction of ~2%. This is the highest record of an ozone fraction in artificially formed hydrates ever reported in the literature. Such an ozone fraction in a formed hydrate may be further increased by increasing the ozone fraction in the feed gas at the cost of the increasing risk of its explosion [10]. The magnitude of the in-hydrate ozone fraction that we achieved in this study seems to be a good compromise between the demand for increasing the ozone content of a formed hydrate and the need for securing operational safety of the hydrate-forming process using an ozone-containing gas mixture as the feed gas.

Besides the magnitude of the ozone fraction in a freshly formed hydrate, the preservability of ozone encaged in the hydrate is of practical importance. The preservation tests performed in this study revealed that the in-hydrate ozone fraction asymptotically decreases with time from its initial value, about 0.8−0.9%, but still remains at about 0.4−0.6% after a 20-day hydrate storage in an aerated atmospheric-pressure condition cooled at −25°C. This finding strongly suggests the practical utility of mixed ozone + oxygen + carbon dioxide hydrates for the industrial, medical and consumer uses of ozone.

ACKNOWLEDGMENTS

We thank Mr. Sanehiro Muromachi, a graduate student at Keio University, for his help in the cage occupancy calculations described in Supporting Information S2.

AUTHOR CONTRIBUTIONS

Conceived and designed the experiments: TN RO YHM. Performed the experiments: TN TK ST. Analyzed the data: TN TK ST. Wrote the paper: YHM.

REFERENCES

1. Sloan ED, Koh CA (2008) Clathrate Hydrates of Natural Gases, 3rd ed. Boca Raton: CRC Press. p. 266.
2. McTurk G, Waller JG (1964) Ozone–carbon tetrachloride double hydrate. Nature 202: 1107. doi: 10.1038/2021107a0
3. Waller JG, McTurk G (1964) Novel ozone inclusion compound. U.K. Patent Specification 961115.
4. Masaoka T, Yamamoto A, Motoi K (2007) Storing method of ozone, method of producing solid material incorporating ozone, food preservation material and food preserving method. Japan Patent Publication 2007–210881. Released online in the Patent and Utility Model Gazette DB of the Japan Patent Office, 23 August 2007. Available: http://www4.ipdl.inpit. go.jp/Tokujitu/tjsogodben.ipdl?N0000 = 115. Accessed 2012 Aug 18.
5. Muromachi S, Ohmura R, Takeya S, Mori YH (2010) Clathrate hydrates for ozone preservation. J Phys Chem B 114: 11430–11435. doi: 10.1021/jp105031n
6. Muromachi S, Nakajima T, Ohmura R, Mori YH (2011) Phase equilibrium for clathrate hydrates formed from an ozone + oxygen gas mixture coexisting with carbon tetrachloride or 1,1-dichloro-1-fluoroethane. Fluid Phase Equilib 305: 145–151. doi: 10.1016/j.fluid.2011.03.020
7. Nakajima T, Akatsu S, Ohmura R, Takeya S, Mori YH (2011) Molecular storage of ozone in a clathrate hydrate formed from an $O_3 + O_2 + CO_2$ gas mixture. Angew Chem Int Ed 50: 10340–10343. doi: 10.1002/anie.201104660
8. Muromachi S, Ohmura R, Mori YH (2012) Phase equilibrium for ozone-containing hydrates formed from an (ozone + oxygen) gas mixture coexisting with gaseous carbon dioxide and liquid water.

J Chem Thermodyn 49: 1–6. doi: 10.1016/j.jct.2012.01.009
9. Mitsuda H, Ominami H, Yamamoto A (1990) Synergistic effect of ozone and carbon dioxide gases for sterilizing food. Proc Japan Acad B 66: 68–72. doi: 10.2183/pjab.66.68
10. Koike K, Nifuku M, Izumi K, Nakamura S, Fujiwara S, et al. (2005) Explosion properties of highly concentrated ozone gas. J Loss Prev Process Ind 18: 465–468. doi: 10.1016/j.jlp.2005.07.020

Chapter 5

Towards a Green Hydrate Inhibitor: Imaging Antifreeze Proteins on Clathrates

Raimond Gordienko[1], Hiroshi Ohno[1,3],
Vinay K. Singh[2], Zongchao Jia[2], John A. Ripmeester[3],
and Virginia K. Walker[1]

[1]Department of Biology, Queen's University, Kingston, Ontario, Canada
[2]Department of Biochemistry, Queen's University, Kingston, Ontario, Canada
[3]Materials Structure and Function Group, National Research Council Canada, Ottawa, Ontario, Canada

ABSTRACT

The formation of hydrate plugs in oil and gas pipelines is a serious industrial problem and recently there has been an increased interest in the use of alternative hydrate inhibitors as substitutes for thermodynamic inhibitors like methanol. We show here that antifreeze proteins (AFPs) possess the ability to modify structure II (sII) tetrahydrofuran (THF) hydrate crystal morphologies by adhering to the hydrate surface and inhibiting growth in a similar fashion to the kinetic inhibitor poly-N-vinylpyrrolidone (PVP). The effects of AFPs on the formation and growth rate of high-pressure sII gas mix hydrate demonstrated that AFPs are superior hydrate inhibitors compared to PVP. These results indicate that AFPs may be suitable for the study of new inhibitor systems and represent an important step towards the development of biologically-based hydrate inhibitors.

INTRODUCTION

Gas hydrates, or clathrates, are ice-like compounds that form when hydrocarbon-based guest molecules are trapped in hydrogen-bonded water cages that form under high pressures and low temperatures [1]. Natural gas hydrates most commonly exist as one of two structures. Small guest molecules such as methane tend to form structure I (sI) hydrates while larger guests like propane form structure II (sII) hydrates [2]. In the laboratory, gas hydrates are conveniently modeled using tetra hydro furan (THF) which is enclathrated at atmospheric pressures [3]. THF hydrate forms cubic sII clathrates, similar to the hydrates that form in pipelines during oil and gas production [4].

Recently, the petroleum industry has been moving into deeper waters which present prime conditions for hydrate growth. Hydrate plugs impede oil and gas flow, resulting in equipment damage as well as hazardous working conditions that can even result in blowouts [5]. Thermodynamic inhibitors such as methanol are one of the most common practical means of controlling hydrate formation [6]. However, as a result of the high costs, flammability and environmental toxicity associated with such inhibitors, there has been a shift towards the less toxic and sometimes cheaper alternative kinetic hydrate inhibitors, which delay nucleation and interfere with crystal growth, as well as antiagglomerants, which act to prevent hydrates from aggregating into larger masses [7], [8].

These concerns have prompted us to investigate the potential inhibitory effects of antifreeze proteins (AFPs) on hydrates. AFPs are a diverse class of proteins that were first identified in fish during the 1950s and have since been found in cold-adapted bacteria, plants and insects[9]–[13]. Despite differences in structure, they have the common ability to adsorb to ice using specific ice-binding faces. AFPs lower the freezing point of water as a result of increased local curvature of growing ice around the adsorbed protein, resulting in a difference between the freezing and melting points, a phenomenon known as thermal hysteresis (TH) [13].

We have previously proposed AFPs as alternative hydrate inhibitors [14]–[15]. Here we visualize fluorescently-tagged AFPs and characterize the effects of these proteins on THF hydrate crystals. We also determine the inhibitory effects of these AFPs on the gas consumption and growth rates of sII natural gas hydrate as part of our efforts towards the development of alternative, biologically-based hydrate inhibitors.

RESULTS

Bioreactor Yields

The AFP cloned from the perennial grass, *Lolium perenne* (Lp), has a low TH while the ocean pout fish Type III AFP is more active [9], [11]. Sequences encoding these AFPs, with or without a green fluorescent protein (GFP) label, were expressed in *E. coli* grown in a bioreactor. Yields were comparable for all recombinant proteins, although optimization of AFP production depended on a variety of conditions including total volume of growth media, level of dissolved oxygen (DO) or OD_{600} at isopropyl β-D-1-thiogalactopyranoside (IPTG) induction as well as temperature during the induction period. Although most of these parameters were kept as constant as possible, it was generally found that lower induction temperatures (≤20°C), yielded the highest amounts of protein. Final purified yields for reactor volumes of 3.5 to 4 L ranged from 180 to 260 mg. Purified recombinant proteins were seen as single bands on a 12% SDS-PAGE gel (Fig. 1). TH values for the AFPs were in the expected ranges at 0.10 ($\pm 8.5 \times 10^{-3}$)°C for LpAFP-GFP, 0.48 ($\pm 2.8 \times 10^{-2}$)°C for Type III AFP-GFP and 0.43 ($\pm 2.8 \times 10^{-2}$)°C for Type III AFP. Control GFP showed no TH. All proteins were tested at ~4 mg/ml.

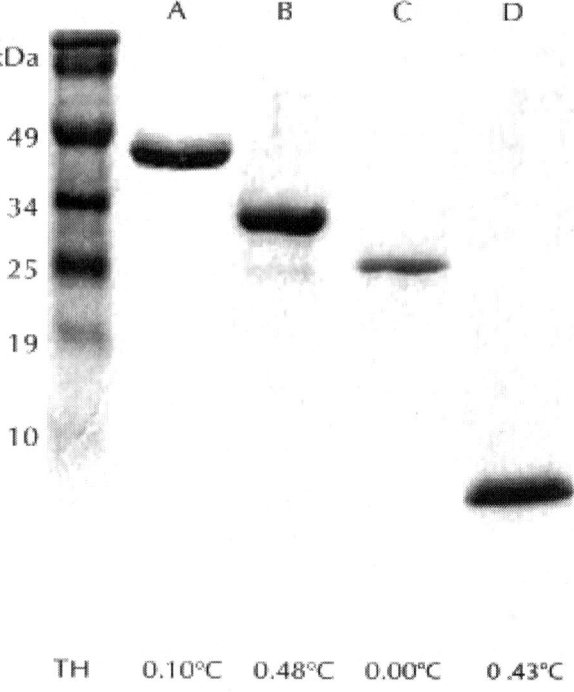

Figure 1: Purification of recombinant proteins. Typical 12% SDS-PAGE analysis of recombinant His-tagged bioreactor-produced proteins including LpAFP-GFP (A), Type III AFP-GFP (B), GFP (C) and Type III AFP (D), purified using Co^{2+}-agarose affinity chromatography. Average TH values are shown below.

Recombinant GFP-Labeled Proteins and Polycrystalline THF Hydrate

Polycrystalline THF hydrate crystals grown in the presence of GFP-tagged AFPs were obviously fluorescent green under UV illumination (Fig. 2). Conversely, when hydrate was grown in the presence of recombinant GFP alone, the hydrates were uniformly dark. In an effort to quantify the adsorption, the hydrates were melted and assayed for adsorbed protein (μmoles of recombinant protein per gram of crystal, Fig. 3). For both purified LpAFP-GFP and Type III AFP-GFP fusion proteins, there was a linear correlation between the amount of protein adsorbed into the growing THF hydrate and the concentration of the protein in the THF solution. At lower concentrations (2 and 4 µM) more Type III AFP-GFP appeared to bind than LpAFP-GFP. At higher concentrations (8 and 16 µM) more LpAFP-GFP adsorbed, to an average of 42% more, than Type III AFP. All differences between the two AFP-GFPs were statistically significant. In contrast, no adsorption of GFP was detected in the crystals at any concentration, with the exception of 2 µM, where an average of 7.9×10^{-5} µmoles/g-crystal was detected. At all other concentrations, differences in the amounts of GFP bound compared to the two AFP-GFPs were statistically significant.

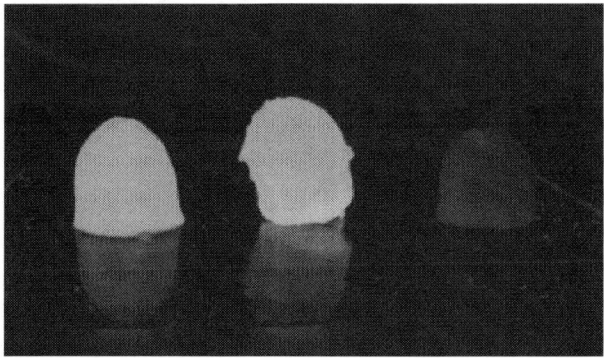

Figure 2: Adsorption of AFPs on THF hydrate. Representative THF hydrate polycrystals fluoresce green under UV light after being grown in solutions containing Type III AFP-GFP (left) and LpAFP-GFP (center). THF hydrate crystals grown in GFP control solutions (right) displayed no fluorescence. Sample diameters were 3–3.5 cm.

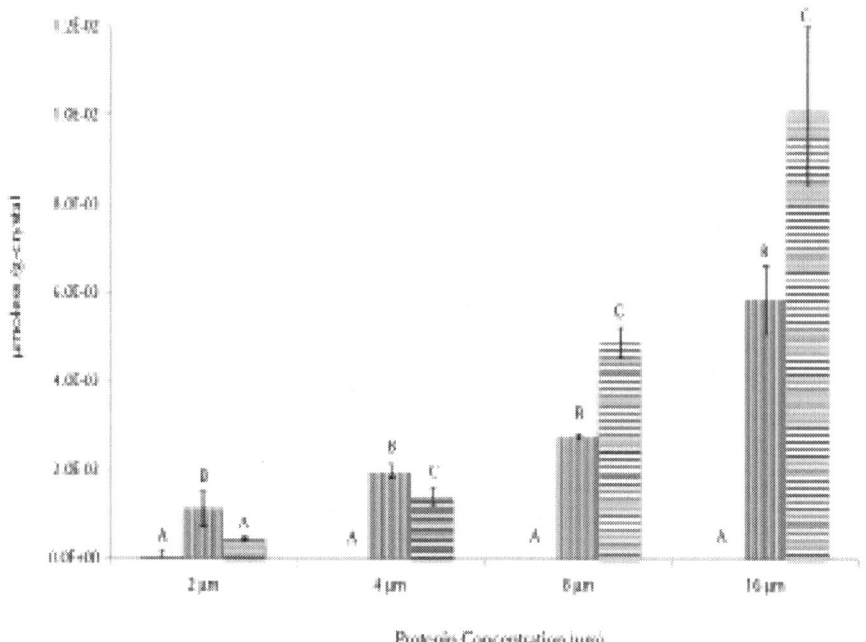

Figure 3: Levels of AFP adsorbed into THF hydrate. Differences in the average μmoles of protein adsorbed per gram of THF hydrate crystal (μmoles/g-crystal) between LpAFP-GFP (horizontal lines), Type III AFP-GFP (vertical lines) and GFP (solid) are plotted as a function of protein concentration (μM). Bars indicate standard deviation. Statistical significance between each data group is indicated by letters A–C, where identical letters indicate no statistical difference.

AFPs and Single THF Hydrate Crystals

Since the fusion proteins adsorbed to polycrystalline THF hydrate, single THF hydrate crystals were slowly grown at low super cooling (3°C) to observe if the presence of the proteins could change crystal morphology. With no proteins, or in the presence of low to moderate concentrations of GFP (Fig. 4A and B), crystals exhibited a cubic octahedral shape that is characteristic of THF hydrate at these conditions [16]. The {111} planes were clearly defined and there were no apparent modifications to crystal shape.

Figure 4: Single THF hydrate crystals. Single crystals grown in solutions of THF/water (A) and 200 μg/ml GFP (B) show no major changes in morphology. Crystals grown in solutions containing 200 μg/ml (30 μM) Type III AFP-GFP (C), 200 μg/ml (4.9 μM) LpAFP-GFP (D) and 15 μg/ml (2.2 μM) Type III AFP (E) show skeletal growth. Crystals with adsorbed Type III AFP-GFP (100 μg/ml; 3.1 μM) (F) and LpAFP-GFP (100 μg/ml; 2.4 μM) (G) show skeletal growth and fluoresce under UV light. Skeletal crystals were grown in 200 μg/ml (20 μM) PVP (H). Crystals (A–H) were grown at 3°C. Control crystals grown in THF/water at high driving force (0°C) show skeletal growth (I). All experiments were done in triplicate and have 4.5–5 cm crystal diameters.

In striking contrast, when single crystals were grown under the same conditions, but in the presence of recombinant AFPs, they exhibited hopper-like or skeletal morphologies (Fig. 4C–G), indicating interference in normal crystal growth habit [17]. At all but the lowest concentration of LpAFP-GFP, depressions were observed on the interior {111} faces. Similar morphologies were seen in crystals grown in solutions of all but the lowest concentration (15 μg/ml, or 0.472 μM) of Type III AFP-GFP. Interestingly, only the non-GFP tagged Type III AFP displayed clear skeletal shaping at the lowest concentration

(15 µg/ml, or 2.2 µM). It was not possible to test this protein at higher concentrations since it precipitated easily in the THF solution. As indicated, recombinant GFP did not appear to alter crystal morphology until tested at very high concentrations (100 and 200 µg/ml, or 4 and 8 µM) and even then showed only one to two shallow face depressions. When the crystals grown in AFP-GFP solutions were observed under UV light, it appeared that the protein had adsorbed uniformly throughout the entire cubic crystal and at every face (Fig. 4F and G). Consistent with the morphological observations, there was no evidence of GFP adsorption under UV light at the low or moderate concentrations used.

In order to compare the effect of a known sII inhibitor on the growth habit of single THF hydrate crystals, crystals were grown in solutions containing poly-N-vinylpyrrolidone (PVP). In all but the lowest concentration of PVP, single crystals were hopper-like and similar to those grown in the presence of AFPs, with the depressions on the {111} faces becoming more pronounced as the concentration of PVP increased (Fig. 4H). It should be noted that all these crystals were grown slowly at temperatures just below the THF hydrate crystallization point. When single crystals were grown at higher driving force, in supercooled conditions at 0°C (Fig. 4I) or at −1.5°C, they exhibited a skeletal morphology like crystals grown in AFPs and PVP.

AFPs and Gas Hydrates

Gas hydrates were formed under high driving force using a natural gas (methane/ethane/propane) mixture so as to produce sII hydrate (Fig. 5). Both average moles of gas consumed (nGas) and average consumption rates followed identical patterns. Because the control samples, without additives, stopped growing at ~25 h (1 500 min), this time was selected as a cutoff point to compare gas consumption and growth rates. On average, the highest level of inhibition was seen in the presence of Type III AFP (with no GFP fusion protein). In the presence of this protein, hydrate formation consumed 6.0 mmol of gas and showed a decrease in growth rate by 18% (a mean of 4.0×10^{-6} nGas/min) compared to control samples which consumed the highest amount of gas at 7.4 mmol and showed the fastest growth rates (4.9×10^{-6} nGas/min). LpAFP-GFP and Type III AFP-GFP demonstrated modestly lower gas consumption (6.6 mmol<gas consumption<6.4 mmol) and decreased growth rates of 10 and 13%, respectively (4.4×10^{-6} nGas/min<growth rate<4.3×10^{-6}

nGas/min). In contrast, the GFP and PVP samples showed the highest levels of gas consumption, except for controls, equally at 6.9 mmol, with a fast growth rate (4.6×10^{-6} nGas/min for both), a 7.1% decrease, compared to the controls.

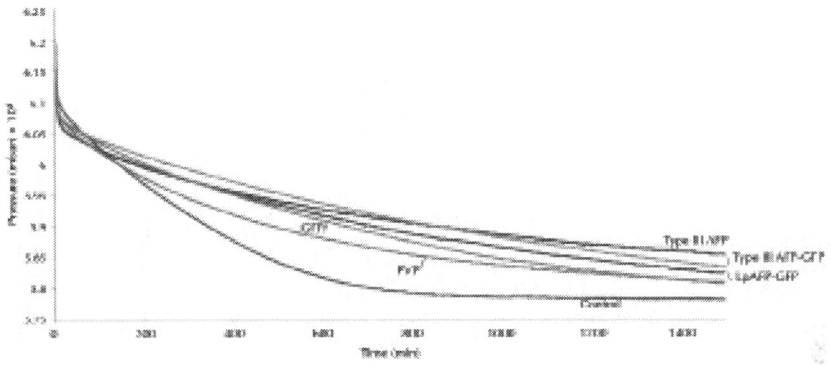

Figure 5: Pressure summary of sII methane/ethane/propane gas hydrate with 0.1 mM additives. Pressure trends plotted against time for Type III AFP, Type III AFP-GFP, LpAFP-GFP, GFP, PVP and control water samples. Absolute pressure drops are proportional to the quantity of moles of gas consumed. Experiments were done in duplicate.

DISCUSSION

The observation that the polycrystalline THF hydrates were strikingly fluorescent after being grown in the presence of recombinant AFP-GFPs (Fig. 2) is irrefutable evidence that these proteins adsorb to sII hydrates. Further substantiation is provided by the morphological changes on single hydrate crystals mediated by these same proteins (Fig. 4). This might not have been predicted *a priori* because although hydrates have an ice-like appearance, their structure is markedly different. Under common conditions, water freezes into hexagonal ice (I_h), taking the form of a hexagonal prism with two basal faces and six rectangular prism faces [18]. Moderately active AFPs such as Type III AFP (and Type III AFP-GFP) have been shown to bind to the ice basal planes, with the low activity LpAFP binding to the prism planes [19], [20]. Thus, evolutionary forces have shaped these AFPs so that they fit securely to these ice surfaces making it more energetically expensive for water molecules to join the growing ice lattice. In contrast, sII hydrates are octahedral with a symmetrical cubic structure and therefore, although similarly flat to ice, present geometrically-distinct, but uniform surfaces for AFP adsorption.

There was no preference by AFPs for any of the 8 identical {111} hydrate faces (Fig. 4F and G), and indeed the resulting hopper-like crystals were similar to those generated in the presence of the commercial kinetic inhibitor, PVP (Fig. 4H). This skeletal growth is believed to occur as a result of the more efficient dissipation of the heat of crystallization on the crystal edges, as opposed to the interior planes, as has been previously described [16]. At higher driving forces (4.4 and 5.9°C subcooling) we demonstrated that these faces were the slowest growing regions of the crystal (Fig. 4I). Since GFP alone did not adsorb to the THF hydrate (Figs. 2–4), we postulate that the inhibition of hydrate growth by AFPs is mediated by the structure of the proteins, and these become adsorbed into the crystal below the equilibrium growth point. Previously, another control protein, cytochrome C, was not seen to effect THF hydrate growth [15]. Curiously perhaps, the plant LpAFP-GFP with low TH activity towards ice but with a higher ice recrystallization inhibition activity than some other AFPs [20], showed 1.6 fold more absorption onto polycrystalline THF hydrate (at least at the higher concentrations) than the moderately ice-active Type III AFP-GFP. However, Type III without the GFP tag appeared to have superior hydrate-shaping, but its hydrate-binding affinity could not be quantified due its vulnerability to THF denaturation at higher concentrations.

These experiments were all conducted with the model THF hydrate, and thus we thought it important to determine if these recombinant proteins could also show inhibition toward natural gas hydrates. Although some fish AFPs and insect AFPs have shown activity as hydrate inhibitors in propane hydrate [21], the GFP fusions used here have never been tested, nor have the tested hydrates been formed using a gas mixture that would be found in a high-pressure oil and gas pipeline [22]. Conditions in pipelines can vary, but deep sea lines are generally at 4°C and ≥1 000 PSI (6.9×10^4 mbar) [7], an environment that we approximated here. These parameters did not seem to have deleterious effects on the proteins because even when the gas hydrates

Single Crystal THF Hydrate

Single THF hydrate crystals were formed at 3°C (1.4°C supercooling) by placing a glass Pasteur pipette, held in place by a central hole punched through a rubber stopper, into a beaker sealed with parafilm containing 80 ml THF solution. Recombinant AFPs, GFP or a commercial hydrate inhibitor, PVP (MW 10 000; Sigma-Aldrich, St. Louis, MO, USA) were added at 15 µg/ml (corresponding to 2.2 µM Type III AFP, 0.47 µM Type III AFP-GFP, 0.37 µM LpAFP-GFP, 0.60 µM GFP and 1.5 µM PVP from here on in), 50 µg/ml (corresponding to molar concentrations of 7.4 µM, 1.6 µM, 1.2 µM, 2.0 µM and 5.0 µM), 100 µg/ml (corresponding to molar concentrations of 15.0 µM, 3.1 µM, 2.4 µM, 4.0 µM and 10.0 µM) and 200 µg/ml (corresponding to molar concentrations of 30.0 µM, 6.3 µM, 4.9 µM, 8.0 µM and 20.0 µM) to observe changes in THF single crystal morphologies. The experiments were performed using wt/vol as opposed to molar concentrations because previous experiments using this technique were done in a similar fashion [16], [17]. In contrast to earlier studies [14], [17], single crystals were never transferred to inhibitor solutions, which can result in additional nucleation sites. To determine the effects of high supercooling rates on crystal structure, some crystals were grown at 0°C and −1.5°C. Crystal growth was initiated inside the pipette by nucleating the solution with a supercooled copper wire placed in dry ice, as described [17]. THF hydrate was formed down the pipette›s decreasing diameter until a single, octahedral crystal emerged from the tip. If more than one crystal was initiated, or the beaker was jarred, the experiment was discontinued, as this resulted in polycrystalline growth. The crystal was grown slowly for approximately 6 h, or until the crystal edges touched the beaker walls. When complete, the crystal was removed from the solution and frozen at −20°C until photographed (Nikon, Coolpix S10).

sII Gas Hydrate

Recombinant proteins Type III AFP-GFP, LpAFP-GFP, Type III AFP and GFP, as well as PVP were mixed at 0.1 mM concentrations with 1 000 Å pore diameter silica gel (Silicycle Chemicals, St. Jean-Baptiste, QC) at a ratio of 1 to 1.3 (v/w) respectively. The inclusion of the gel allowed more predictable and consistent nucleation times [24]. It is composed

Polycrystalline THF Hydrate

Polycrystalline THF hydrate was grown on a hollow brass finger connected to a temperature-programmable 1197P water bath (VWR International, Mississauga, ON) filled with a water/ethylene glycol mixture (at 3:1 v/v). A solution of THF/water (1:3.34, v/v; 80 ml) was cooled to 4°C and poured into a 100 ml glass beaker, placed in an insulated box. The cooled brass finger (2.5°C) was submerged into the THF/water solution, which was stirred with a magnetic stir bar. To expedite hydrate formation, a small piece of THF hydrate crystal was added into the solution, which helped nucleate a thin layer of hydrate on the surface of the brass finger, a process known as seeding. When the THF hydrate uniformly covered the finger, the beaker was removed and replaced with another containing 80 ml of the test solution, and sealed with parafilm to prevent the evaporation of THF.

Triplicate test solutions of consisted of THF/water (1:3.34, v/v; 80 ml) containing 2 µM, 4 µM, 8 µM and 16 µM of purified recombinant LpAFP-GFP, Type III AFP-GFP and GFP. THF polycrystals were grown for 8 h with the ethylene/glycol bath set to an initial temperature of 0°C and a final temperature of −6.5°C, with a 0.5°C drop every 30 min. The temperature of the THF/water/protein solution surrounding the growing crystal remained at approximately 4.9°C for the duration of the experiment. At the conclusion of the growth period, the beaker was removed and the adhered polycrystal was wrapped in aluminum foil and placed in ice to prevent melting. The bath temperature was then increased to 5.5°C, heating the brass finger and allowing the crystal to be collected.

After washing the crystal with 30 ml of distilled water (<4°C) to remove any residual solution, it was weighed and examined under UV light (midpoint wavelength = 302 nm) at 4°C and photographed. After melting the polycrystals at room temperature, the solution was concentrated to 2 ml with a 15 ml concentrating tube (Millipore, Billevica, MA, USA), and the protein concentration was determined as above. The amount of recombinant protein adsorbed per gram of hydrate crystal (denoted as $mol/g_{crystal}$) was calculated and a Tukey honestly significant difference (HSD) test was used for statistical comparisons ($\alpha = 0.05$, $q^* = 3.07$).

with a poly(His)$_6$ sequence located on the C-terminal of the expressed proteins to facilitate purification.

THF (≥99.5%, Sigma-Aldrich, St. Louis, MO, USA) was mixed with distilled water or water containing solutions of proteins or other additives, at a 1:15 molar ratio or 1: 3.34 (v/v) as previously described [3]. The observed crystallization temperature was the same as published values (≤4.4°C) [15].

Bioreactor: Production and Purification

Recombinant *E. coli* BL21 cells were grown in a New Brunswick BioFlow 110 bioreactor (Edison, NJ, USA) using Luria-Bertani (LB) media enriched with 5 g/L yeast extract, 6 g/L glucose, 12 g/L Na$_2$HPO$_4$, 6 g/L K$_2$HPO$_4$, 2 g/L NH$_4$Cl, 0.022 g/L CaCl$_2$ and 0.482 g/L MgSO$_4$ at 37°C, pH 7, 650 RPM agitation and 4 L/min air flow, until the culture reached OD$_{600}$ = 8 or the DO levels were at a minimum (0–20%). The culture temperature was then decreased to 20°C, and recombinant protein expression was induced with 1 mM IPTG for 16 h. The cells were harvested by centrifugation (6 000×g, 4°C, for 20 min) and resuspended in cold lysis buffer (20 mM Tris, 500 mM NaCl, pH 7, containing an EDTA-free protease inhibitor cocktail (Mini pills; Roche, Manheim, Germany).

The pelleted cells were then lysed using a Branson sonicator for 4 min with 1 min cooling on ice and centrifuged (16 000×g, 4°C for 40 min) to remove solid cell debris. The supernatant was then mixed with cobalt-based agarose Talon Metal Affinity Resin (BD Bioscience, Mountain View, CA, USA) and incubated for 1.5 h at 4°C with mild shaking, allowing the poly(His)-tagged proteins to bind to the resin. The supernatant/resin mixture was then loaded onto a 100 ml column and the resin was washed twice with wash buffer (20 mM Tris, 500 mM NaCl, pH 7), once with wash buffer containing 5 mM imidazole, followed by 10 mM imidazole and finally eluted with 20 mM and 250 mM imidazole. The protein was visualized on a 12% SDS-PAGE gel stained with Coomassie Brilliant Blue R-250 to assess purity. Protein concentration was determined using dye-binding via bicinchurinic acid assay (BCA Protein Assay Kit; Pierce, Rockford, IL, USA). Thermal hysteresis (TH) was measured as previously described [23] with a nanolitre osmometer at protein concentrations of approximately 4 mg/ml (Clifton Technical Physics, Hartford, NY, USA).

were thawed and the proteins used a second time, inhibition activity was still observed (not shown). All of the investigated AFPs showed hydrate inhibition as determined by gas uptake assessments. Similar to the observations on the single THF crystals, LpAFP-GFP and Type III AFP-GFP showed hydrate inhibition that was modestly higher than the chemical inhibitor PVP. GFP showed little inhibition, possibly only due to a colligative effect since, as indicated, no incorporation into THF hydrate was seen. Of the additives tested, Type III AFP was demonstrably superior with an overall 18% decrease in gas hydrate formation, validating again the observations made with this protein on THF single hydrate crystals.

In conclusion, we have demonstrated for the first time that AFPs irreversibly adsorb to sII hydrate surfaces and we speculate that they act as inhibitors by binding to the {111} faces of the these symmetrical cubic crystals. We further consider that the identified ice-binding residues of these proteins may not be identical to the residues that interact with the hydrate surface, but the way is now clear for such an investigation. In addition, these experiments have established that AFPs are suitable models for understanding hydrate-inhibitor reactions and offer the prospect that these proteins, or their modified cognates, will be useful as new and more effective biologically-based hydrate inhibitors.

MATERIALS AND METHODS

Bacterial Strains and THF Solutions

Sequences encoding the AFPs were expressed in *E. coli* BL21 cells. The plasmids used included the pET-24 vector (Novagen) for the expression of Type III AFP (7 kDa) and LpAFP-GFP (41 kDa), and pET-20b (Novagen) for Type III AFP-GFP (32 kDa). A plasmid encoding a control GFP (25 kDa) without an AFP sequence, was made by the amplification of GFP from the pET-20 vector expressing the Type III AFP-GFP sequence using primers with a 5'-NdeI site and a 3'-HindIII site. The amplified GFP was then inserted into a pET24a vector between its NdeI/HindIII sites, followed by subsequent transformation of BL21 cells. The insert was verified by sequencing. All of the recombinant proteins were tagged of an organic form of silicon and not amorphous silica, to which other AFPs have been reported to adsorb [25]. Briefly, the silica sand/additive mix was added into a stainless steel cell. After the sample system was immersed into a water bath set to 25°C, the sample was purged twice by a natural gas mix consisting of 2% propane, 5% ethane and 93% methane (Linde Canada, Mississauga, ON) at approximately 250 PSI (1.7×10^4 mbar), and then was pressurized to 1000 PSI (6.9×10^4 mbar) with the same gas mixture. Hydrate nucleation was initiated by transferring the pressurized cell into another water bath set to 0.5°C. Hydrate formation was monitored by a sudden drop in pressure (recorded using an Omega DAQPRO-5300; Fourier Systems, Fairfield, CT, USA). Moles of gas consumed (nGas), indicating the amount of hydrate formed, were calculated as previously described [26] with a minor modification.

$$nGas = V\,(P/zRT)_0 - V\,(P/zRT)_t$$

This relationship allows the determination of the difference between moles of gas at time $t = 0$ and the number of moles of gas at time t, where the V is the volume of the gas phase, or the cell volume minus the sample volume, P is pressure, R is the gas constant, T is temperature and z is the compressibility factor, calculated by Pitzer's correlations [27] and assuming no change in gas phase volume coincident with hydrate formation. Additionally, the rate of change of gas consumption, denoted as nGas/min (indicative of the rate of hydrate growth), was calculated by determining the average of the slopes of all nGas data points (slope = ΔnGas/Δtime). Experiments were performed in duplicate.

ACKNOWLEDGMENTS

We thank Adam Middleton, Sherry Gauthier and Peter L. Davies for providing the AFP constructs and Zhongqin (Suzy) Wu for aiding in the construction of the recombinant GFP plasmid. We also acknowledge the editor and referees for their suggestions.

AUTHOR CONTRIBUTIONS

Conceived and designed the experiments: RG HO VKW. Performed the experiments: RG HO. Analyzed the data: RG HO. Contributed reagents/materials/analysis tools: VKS ZJ JAR VKW. Wrote the paper: RG.

REFERENCES

1. Kvenvolden KA (1993) Gas hydrates- geological perspective and global change. Rev Geophys 31: 173–187.
2. Sloan ED Jr (2003) Fundamental principles and applications of natural gas hydrates. Nature 426: 353–363.
3. Makogon YT, Larsen R, Knight CA, Sloan ED (1997) Melt growth of tetrahydrofuran clathrate hydrate and its inhibition: method and first results. J Cryst Growth 179: 258–262.
4. Davies SR, Selim MS, Sloan ED, Bollavaram P, Peters DJ (2006) Hydrate plug dissociation. AIChE J 52: 4016.
5. Mehta A, Walsh J, Lorimer S (2006) Hydrate challenges in deep water production and operation. Ann NY Acad Sci 912: 366–373.
6. Koh CA, Westacott RE, Zhang W, Hirachand K, Creek JL, et al. (2002) Mechanisms of gas hydrate formation and inhibition. Fluid Phase Equilib 30: 143–151.
7. Lederhos JP, Long JP, Sum AK, Christiansen RL, Sloan ED Jr (1996) Effective kinetic inhibitors for natural gas hydrates. Chem Eng Sci 51: 1221–1229.
8. Huo Z, Freer E, Lamar M, Sannigrahi B, Knauss DM, et al. (2001) Hydrate plug prevention by anti-agglomeration. Chem Eng Sci 56: 4979–4991.
9. Gordon MS, Amdur BH, Scholander PF (1962) Freezing resistance in some northern fishes. Biol Bull 122: 52–62.
10. Gilbert JA, Davies PL, Laybourn-Parry J (2005) A hyperactive, Ca^{2+}-dependent antifreeze protein in an Antarctic bacterium. FEMS Microb Lett 245: 67–72.

11. Middleton AJ, Brown AM, Davies PL, Walker VK (2009) Identification of the ice-binding face of a plant antifreeze protein. FEBS Lett 583: 815–819.
12. Tyshenko MG, Doucet D, Davies PL, Walker VK (1997) The antifreeze potential of the spruce budworm thermal hysteresis protein. Nature 15: 887–890.
13. Barrett J (2001) Thermal hysteresis proteins. Int J Biochem Cell Biol 33: 105–117.
14. Zeng H, Wilson LD, Walker VK, Ripmeester JA (2003) The inhibition of tetrahydrofuran clathrate-hydrate formation with antifreeze proteins. Can J Phys 81: 17–24.
15. Zeng H, Wilson LD, Walker VK, Ripmeester JA (2006) Effect of antifreeze proteins on the nucleation, growth, and the memory effect during tetrahydrofuran clathrate hydrate formation. JACS 128: 2844–2850.
16. Larsen R, Knight CA, Sloan ED Jr (1998) Clathrate hydrate growth and inhibition. Fluid Phase Equilib 150: 353–360.
17. Knight CA, Rider K (2002) Free-growth forms of tetrahydrofuran clathrate hydrate crystals from the melt: plates and needles from a fast-growing vicinal cubic crystal. Philos Mag A 82: 1609–1632.
18. Materer N, Starke U, Barbieri A, Van Hove MA, Somorjai GA, et al. (1995) Molecular surface structure of a low-temperature ice Ih (0001) crystal J Phys Chem 99: 6267–6269.
19. Scotter AJ, Marshall CB, Graham LA, Gilbert JA, Garnham CP, et al. (2006) The basis for hyperactivity of antifreeze proteins. Cryobiology 53: 229–239.
20. Pudney PDA, Buckley SL, Sidebottom CM, Twigg SN, Sevilla M-P, et al. (2002) The physico-chemical characterization of a boiling stable antifreeze protein from a perennial grass (*Lolium perenne*). Arch Biochem Biophys 410: 238–245.
21. Zeng H, Moudrakovski IL, Ripmeester JA, Walker VK (2006) Effect of antifreeze protein on nucleation, growth and memory of gas hydrates. AIChE J 54: 3304–3309.
22. Kennedy JL (1993) Oil and gas pipeline fundamentals, 2nd Ed. Tulsa, OK,. USA: Pennwell Corp.

23. Chakrabartty A, Hew CL (1991) The effect of enhanced α-helicity on the activity of a winter flounder antifreeze polypeptide. Eur J Biochem 202: 1057–1063.
24. Seo Y, Ripmeester JA, Lee J, Lee H (2005) Efficient recovery of CO_2 from flue gas by clathrate hydrate formation in porous silica gels. Environ Sci Technol 39: 2315–2319.
25. Zeng H, Walker VK, Ripmeester JA (2007) Approaches to the design of better low-dosage gas hydrate inhibitors. Angew Chem 119: 5498–5500.
26. Lee JD, Englezos P (2005) Enhancement of the performance of gas hydrate kinetic inhibitors with polyethylene oxide. Chem Eng Sci 60: 5323–5330.
27. Smith JM, Van Ness HC, Abbott MB (2001) Introduction to Chemical Engineering Thermodynamics. New York: McGraw-Hill.

Comparison and Analysis of Zinc and Cobalt-Based Systems as Catalytic Entities for the Hydration of Carbon Dioxide

Edmond Y. Lau, Sergio E. Wong, Sarah E. Baker, Jane P. Bearinger, Lucas Koziol, Carlos A. Valdez, Joseph H. Satcher Jr, Roger D. Aines, and Felice C. Lightstone

Physical and Life Sciences Directorate, Lawrence Livermore National Laboratory, Livermore, California, United States of America

ABSTRACT

In nature, the zinc metalloenzyme carbonic anhydrase II (CAII) efficiently catalyzes the conversion of carbon dioxide (CO_2) to bicarbonate under physiological conditions. Many research efforts have been directed towards the development of small molecule mimetics that can facilitate this process and thus have a beneficial environmental impact, but these efforts have met very limited success. Herein, we undertook quantum mechanical calculations of four mimetics, 1,5,9-triazacyclododedacane, 1,4,7,10-tetraazacyclododedacane, tris(4,5-dimethyl-2-imidazolyl)phosphine, and tris(2-benzimidazolylmethyl)amine, in their complexed form either with the Zn^{2+} or the Co^{2+} ion and studied their reaction coordinate for CO_2 hydration. These calculations demonstrated that the ability of the complex to maintain a tetrahedral geometry and bind bicarbonate in a unidentate manner were vital for the hydration reaction to proceed favorably. Furthermore, these calculations show that the catalytic activity of the examined zinc complexes was insensitive to coordination states for zinc, while coordination states above four were found to have an unfavorable effect on product release for the cobalt counterparts.

INTRODUCTION

In recent years a growing awareness of carbon dioxide atmospheric levels sparked interest in developing rapid methods for the capture and sequestration of the gas from industrial gas streams [1]. Most industrial separation processes for CO_2 involve a liquid in which the dissolved gas ionizes under highly basic conditions, leading to its full dissolution and concomitant adsorption into the medium [2]. The rate-limiting step in such processes is well known to be the formation of carbonic acid. The slow kinetics nature of this reaction also hinders the uptake of CO_2 in the ocean, and it is the underlying cause of the significant mass transfer limitation at the water's surface [3]. This mass transfer limitation also applies to industrial gas separations [4],[5], [6] and results in overall decreases by a factor of 1000-fold over that which could be obtained, if the hydration of the CO_2 was not the rate-limiting step. Accelerating such processes through the use of catalysts or enzymes would permit smaller and less expensive separation processes to remove CO_2 from industrial gas emissions [7] and could conceivably be fast enough to permit removal of CO_2 from the atmosphere in processes of the type envisioned by Elliot *et al* [8] and Keith *et al* [9].

In biological systems the reversible hydration of CO_2 to bicarbonate is carried out with formidable efficiency by the zinc metalloenzyme, carbonic anhydrase (CA) [10]. In humans, carbonic anhydrase II (CAII, EC 4.2.1.1) is the most efficient isoform exhibiting activity that approaches diffusion limited kinetics. The reaction is catalyzed by a zinc-hydroxide containing center that is formed upon deprotonation of a water molecule coordinated to the active site's zinc (Zn-OH_2, pK_a ~7) [11]. The reaction mechanism, which follows ping-pong kinetics, occurs via two independent steps [10], [12]. In step one, the zinc-hydroxide in the active site of CA nucleophilically attacks CO_2 to form a Zn^{2+} bound bicarbonate intermediate whose reaction with water results in the expulsion of bicarbonate.

$$\text{E-OH}^- + CO_2 \leftrightarrow \text{E-HCO}_3^- \xleftrightarrow{H_2O} \text{E-}H_2O + HCO_3^-$$

(1)

In the second step, the zinc bound water is deprotonated by a nearby histidine (His64 in human CAII) regenerating the catalytic species while the proton is shuttled into the bulk solvent.

$$\text{E-}H_2O + B \leftrightarrow \text{E-OH}^- + BH^+$$

(2)

Deprotonation of the water is the rate-limiting step in carbonic anhydrase [12]. The extremely high hydration turnover of CO_2 by human CAII is ~10^6 sec^{-1} at pH 9 and 25°C [11], [13]. The reverse reaction, dehydration of bicarbonate occurs when the solution pH is below 7.

The X-ray crystal structures of different CAs have been solved and studied in great detail [10]. Crystallographic studies of human CAII show that the enzyme is a monomeric protein consisting of 260 residues. The funnel-shaped appearance of the active site ends with the zinc metal located in its very interior and tetrahedrally coordinated by three histidines (His94, His96, and His119) and a water/hydroxide molecule [14], [15]. The active site can be divided into a hydrophobic half (Val121, Val143, Leu198, and Trp209) necessary for CO_2 binding and a hydrophilic half (His64 and Thr199), possessing residues and water molecules intimately involved in an intricate hydrogen bonding network for efficient proton shuttling during the last step of the catalysis.

Other divalent metals (Cu^{2+}, Hg^{2+}, Fe^{2+}, Cd^{2+}, Ni^{2+}, Co^{2+} and Mn^{2+}) [16] can bind to CAII, but only Co^{2+} has wild-type catalytic efficiency ($k_{cat}/K_m = 8.7\times10^7$ $M^{-1}s^{-1}$ for Zn^{2+} vs 8.8×10^7 $M^{-1}s^{-1}$ for Co^{2+}), although the individual k_{cat} and K_m values for CAII differ when binding the two metal ions [17]. Due to the lack of spectroscopic signatures by the Zn^{2+}ion, its divalent counterpart Co^{2+} has played an important role in studying CA, not only because it also utilizes metal-hydroxide catalysis and retains near wild-type activity but because it also acts as a spectroscopically active tag [18].

Despite the merits of CAII, current research into the use of carbonic anhydrase for industrial CO_2 capture has faced significant challenges mainly due to the challenging task of producing a viable enzyme for the rigorous demands encountered in industrial processes. Trachtenberg et al [7], [19] have reported the use of a membrane-countercurrent system originally designed for spacecraft use, and Bhattacharya et al [20] developed a spray system containing carbonic anhydrase. Azari and Nemat-Gorgani [21] examined means of using the reversible unfolding of the enzyme, caused by heat, to attach it to more sturdy substrates for industrial use. Lastly, Yan et al [22] incorporated single carbonic anhydrase molecules in a spherical nanogel, resulting in improved temperature stability of the enzyme with only moderate loss of activity. Another route of exploration and one that has been undertaken by several groups is to synthesize small molecules capable of mimicking the enzyme's catalytic property. Creating such mimetics requires incorporating key structural features from the enzyme scaffold and avoiding possible degradation mechanisms of the catalytic center. Fortunately, CA mimetics were developed to study the enzyme's reaction mechanism, and several examples of small molecule CA mimetics exist [23]. In the small molecule mimetics developed to date, the most prominent features of the enzyme's catalytic site, namely the nitrogen atoms belonging to the histidine side chains, have been used as guiding factors in their design. These nitrogen atoms may be part of an imidazole group [24], such as tris(4,5-di-n-propyl-2-imidazolyl) phosphineor nitrilotris(2-benimidazolylmethyl-6-sulfonate),or simply secondary amines, as in to case of 1,5,9-triazacyclododecane [25] or 1,4,7,10-tetraazacyclododecane [26] which chelate the a metal ion to form the catalytic species (Figure 1). These four small molecule mimetics when chelated with Zn^{2+} have been reported to catalyze the hydration of CO_2, although with a more modest catalytic activity compared to the enzyme.

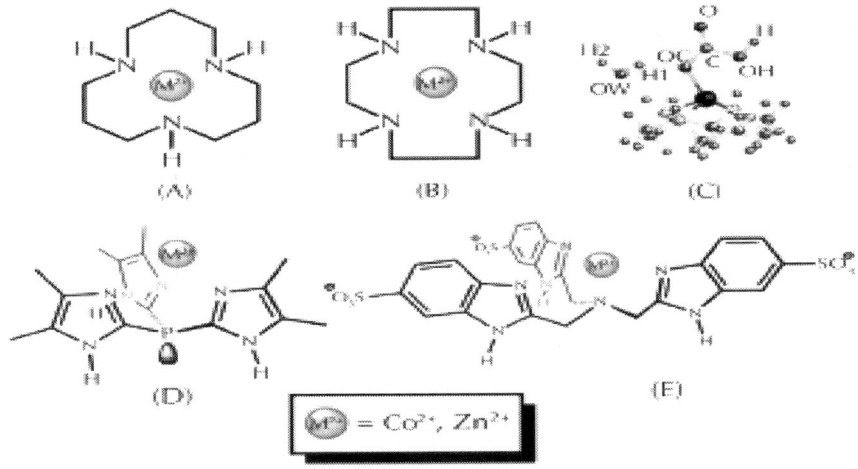

Figure 1: Molecular structures of 1,5,9-triazacyclododedacane (A) and 1,4,7,10-tetraazacyclododedacane (B) which are denoted N3 and N4 in the text, respectively. Panel (C) shows how the atoms are denoted in the text. The structures of tris(4,5-dimethyl-2-imidazolyl)phosphine (D) and tris(2-benzimidazlylmethyl)amine (E) are denoted Ph and Ben in the text, respectively.

In this *ab initio* study, we have examined carbon dioxide hydration as catalyzed by 1,4,7,10-tetraazacyclododedacane (N4), 1,5,9-triazacyclododedacane (N3), tris(4,5-dimethyl-2-imidazolyl)phosphine (Ph), and tris(2-benzimidazolylmethyl)amine (Ben), chelating both Zn^{2+} and Co^{2+}, to investigate the reaction mechanism of these two metals and determine the cause for the difference in activity seen in human CAII.

METHODS

Quantum Mechanical Calculations

The hydration reaction of CO_2 catalyzed by N3, N4, Ph, and Ben, chelating Zn^{2+} and Co^{2+}, was investigated using quantum mechanical calculations. All calculations were carried out using the programs Gaussian03 [27] and Gaussian 09 [28]. Geometry optimizations were performed at the B3LYP/6-311+G(d) level of theory [29], [30].

The catalytically active form of cobalt in carbonic anhydrase is experimentally known to be a high-spin quartet (S = 3/2) [31]. Ground state calculations of these mimetics containing low-spin (S = 1/2) Co^{2+} were consistently higher in energy than the high-spin system. Thus, calculations on the cobalt-containing mimetics were carried out with a fixed quartet multiplicity. The stability of the wavefunction was determined by using the STABLE option within Gaussian. The counterpoise method of Boys and Bernardi was used to account for basis set superposition error (BSSE) [32]. To test the suitability of the B3LYP functional for these calculations, full optimizations ofN4-metal reaction were performed, using a recent functional (MPWLYP1M/6-311+G(d)) that has been successfully used for several organometallics [33]. Harmonic frequency calculations were performed on all the structures to characterize the stationary points. Transition states were characterized by a single imaginary frequency, and their values are provided in. The calculated zero-point energies (ZPE) were not scaled. To investigate the effects of solvation on the hydration reaction, single point calculations using the gas-phase geometries were carried out, using a conductor-like polarizable continuum model (CPCM) [34]to approximate solvent effects (water, = 78.4). It has been previously shown that the solvation free energies from single point PCM calculations, using gas-phase geometries from density functional calculations, are in reasonable agreement with values obtained from full optimizations [35], [36]. All solvation calculations used the simple united atom topological model (UA0) [37], using UFF radii [38]. The gas phase zero point energies were included in the solvation calculations. Natural population analysis was performed on the optimized structures to assess the charge distributions on these complexes [39].

Synthesis

Tris(6-Sulfobenzimidazolylmethyl)Amine (Sulfonated-Ben).

The ligand was synthesized following a previously published protocol for the synthesis of tris-benzimidazole-based compounds [40]. Thus, 4-sulfo-1,2-diaminobenzene [41] (4.0 g, 21.2 mmol) was transferred into a 250 mL round bottom flask equipped with a large stir bar. The solid was made into a suspension with the addition of ethylene glycol (120 mL). To the suspension, nitrilotriacetic acid (1.13 g, 5.89 mmol) was added in one portion, the flask equipped with a condenser (set with water at 10°C) and the resulting mixture was heated to 210°C, using a sand bath overnight. After 18 hours of heating, the flask was removed from the sand bath, and the black-colored reaction mixture was allowed to cool down to ambient temperature. The mixture was subsequently poured into a 1000 mL Erlenmeyer flask containing ice water (300 mL) in small portions with constant swirling. The grey precipitate was collected using vacuum filtration and washed copiously with cold, deionized water (5×50 mL) and dried under vacuum to afford the title compound (2.90 g, 74%). The sodium salt of the ligand was obtained by reacting the ligand (1.0 g, 1.5 mmol) with NaOH (180 mg, 4.5 mmol, 3.05 equiv. to ligand) in deionized water (10 mL). The water was evaporated under reduced pressure to yield a light grey solid (1.03 g, 97%). ^{1}H NMR (600 MHz, D_2O) 7.94 (s, 3H), 7.53 (d, J = 8.5, 3H), 7.35 (d, J = 8.5, 3H), 4.07 (s, 6H); ^{13}C NMR (150 MHz, D_2O) 155.4, 139.6, 137.9, 136.5, 119.7, 114.6, 112.8, 53.2; Anal. ($C_{24}H_{18}N_7Na_3O_9S_3 \bullet H_2O$) C, 39.40; H, 2.76; N, 13.40; Found: C, 39.32; H, 3.08; N, 13.49. The characterization of the zinc complex and the protocols for the kinetic analysis for sulfonated-Ben can be found in.

RESULTS AND DISCUSSION

Our *ab initio* calculations investigated the hydration of CO_2 catalyzed by the Zn^{2+} containing catalysts (Figure 1). The hydration of CO_2 by carbonic anhydrase and mimetics is believed to follow the same reaction pathway. Thus, the catalytic cycle begins with nucleophilic attack on the CO_2 by the zinc-hydroxide species to form Zn^{2+}-bicarbonate intermediate followed by displacement of the bicarbonate from Zn^{2+} by water; the water then loses a proton to regenerate the catalysis. Cobalt substituted carbonic anhydrase also utilizes the above metal-hydroxide reaction mechanism for CO_2 hydration, but it has ~50% of the catalytic activity exhibited by the wild-type enzyme [17]. To gain insight in the possible fine differences between the two metals, chelators containing Co^{2+} were also studied. The cobalt complexes are assumed to be in alkaline conditions which favor tetrahedral geometries and share similar characteristics to the zinc complexes [42],

[43]. There have been several *ab initio* studies on the hydration of CO_2 by CAII [44], [45], [46], [47], [48], [49] but a thorough comparative study between Zn^{2+} and other metals ions within CA have not been as widely studied [50], [51], [52]. Additionally, tris (4,5-di-n-propyl-2-imidazolyl)phosphinecatalyzes the hydration of CO_2, but the non-catalytic tris(4,5-dimethyl-2-imidazolyl)phosphine (Ph) was chosen for computational tractability and can provide insights into the reaction mechanism.

Nucleophilic Attack of CO_2

The first step of the catalyzed hydrolysis of CO_2 in the gas-phase is formation of an encounter complex (EC) between the separated reactants (Figure 2). The EC is formed when CO_2 interacts weakly with one of the amine hydrogens in the ring structure of the macrocycles N3and N4 (Figure 3). Due to the lack of N-H moieties around the catalytic OH^- group in the Ph andBen ligands, only van der Waals complexes were formed with CO_2. The stabilization energy is approximately −1 to −4 kcal/mole for each of the Zn^{2+} and Co^{2+} encounter complexes relative to the separated reactants (Figures 4). The N3 and N4 ECs were found to have greater stabilization energies than the Ph and Ben ECs. The amine hydrogen to CO_2 oxygen distances were measured to be 2.071 and 2.124 Å for N3-Zn and N4-Zn, respectively, while the Co^{2+}complexes had similar distances of 2.090 Å (N3-Co) and 2.322 Å (N4-Co). Additionally, the angle formed by the CO_2 oxygen with the hydrogen and nitrogen of the amine (O•••H-N) of N3and N4 shows that the CO_2 is likely not forming a strong hydrogen bond. Both N3 M^{2+}-complexes have angles close to 180°, whereas the N4 complexes possess angles of 137° for Co^{2+} and 158° for Zn^{2+}. The distances between the CO_2 carbon and the M^{2+}-hydroxide oxygen were 2.674 (N3) and 2.640 Å (N4) for the Zn^{2+} EC structures, while the Co^{2+} complexes had slightly longer distances. The value obtained for N3-Zn is similar to that obtained by Brauer et al. at the HF/6-311+G(d) level of theory, while our N4-Zn value is almost 0.1 Å shorter than Brauer's value [53]. Calculations using the B3LYP or MPWLYP1M functional provide similar results for both M^{2+}-complexes. Only minor differences in the energies and geometries were found for the N4 reaction with either Zn^{2+} or Co^{2+} between these fully optimized calculations.

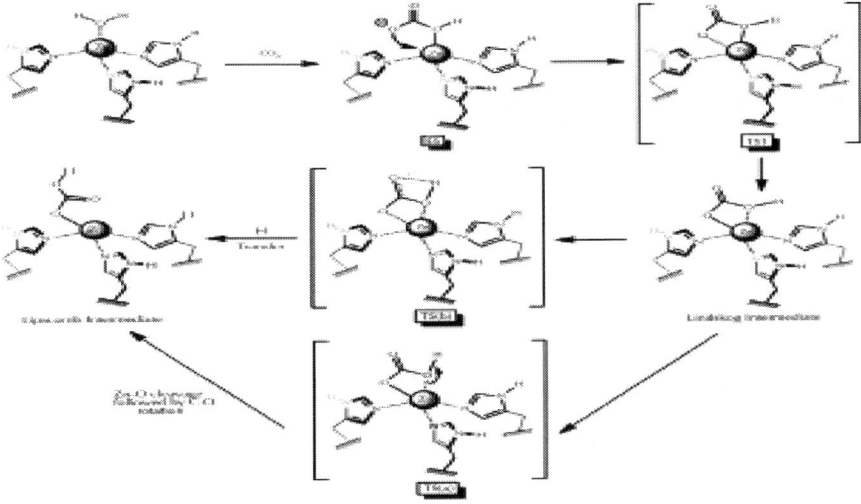

Figure 2: Schematic of the nucleophilic attack of CO_2 by the mimetics and the resulting geometries of the Lindskog and Lipscomb intermediates.

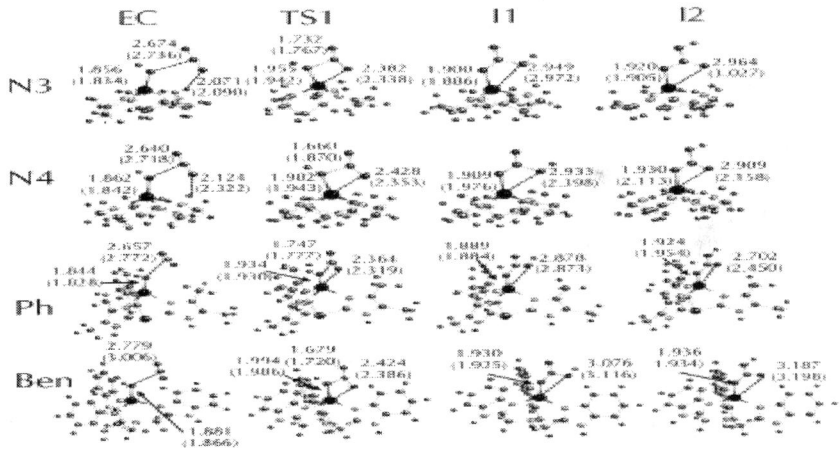

Figure 3: Calculated structures for the zinc complexes of N3, N4, Ph, and Ben complexes during the nucleophilic attack portion of the hydration reaction of CO_2 when forming the encounter complex (EC), first transition state (TS1), Lindskog intermediate (I1), and Lipscomb intermediate (I2). Distances are listed in angstroms and values in the parenthesis are the corresponding distances for the cobalt complexes.

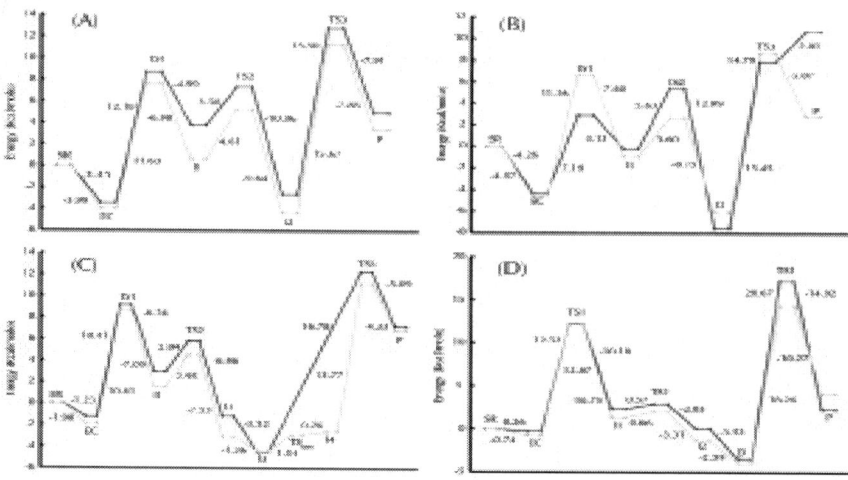

Figure 4: Relative energy of the calculated stationary points for N3 in Panel (A), N4 in Panel (B), Ph in Panel (C), and Ben in Panel (D) along the reaction coordinate relative to the separated reactants (SR). The energies for the zinc complexes are represented by the gray line and the cobalt complexes by the black line.

The calculated distances for the EC structures are in reasonable agreement with a recently solved crystal structure of human carbonic anhydrase (HCAII) with CO_2 [54]. In this crystal structure (PDB ID 3D92), the CO_2 carbon to Zn^{2+}-hydroxide oxygen distance is 2.791 Å. The CO_2 is bound in a hydrophobic pocket within HCAII, and one of its oxygens interacts with the amide backbone nitrogen of Thr199 (3.493 Å). Interestingly, the same study also showed that a metal ion is not even necessary for CO_2 to bind in the correct location in the HCAII active site. Although Ph-metal and Ben-metal complexes lack an N-H group to stabilize the CO_2 around the ligands, the distance from the M^{2+}-hydroxide oxygen to the CO_2 carbon was comparable to the N3 and N4 ligands, even though their stabilization energy is smaller (Figure 3). Natural population analysis (NPA) of the complexes show that there is little charge difference between Zn^{2+} in N3 or N4, a finding that is consistent with the work of Brauer *et al* [53]. Two additional pieces of information obtained from NPA are: (1) there exists some charge polarization occurring in the CO_2 molecule from its interaction with the amine and (2) the Zn^{2+}-hydroxide is more nucleophilic in nature than its Co^{2+} counterpart.

The first transition state (TS1) in the hydration reaction is formed when the distance between the M^{2+}-hydroxide oxygen to the carbon of CO_2 falls below 2 Å. In the Co^{2+} complexes, the distances for the N3-Co, N4-Co, Ph-Co, and Ben-Co transition states were 1.767, 1.870, 1.777, and 1.720 Å, respectively (Figure 3). The N3-Zn complex has a similar transition state distance to N3-Co of 1.732 Å, but the N4-Zn structure has a much shorter distance of 1.660 Å relative to N4-Co. The reaction barriers are similar for N3-Zn, N4-Zn, and N3-Co (~12 kcal/mole), while the reaction barrier for N4-Co is significantly lower at 7.2 kcal/mole (Figure 4B). The difference in energy is due to an earlier transition state for N4-Co than found in the other complexes. Although the CO_2 to M^{2+}-hydroxide distance in N4-Co is longer than in N4-Zn, the oxygen in CO_2 shows greater coordination to Co^{2+} (2.353 Å) relative to Zn^{2+} (2.428 Å). The TS1 geometries for both Ph-metal structures are similar. The distance between the hydroxyl oxygen and carbon of CO_2 is 1.747 for Ph-Zn. Both Ben-metal structures had late transition states that lead to high activation barriers for the final formation of bicarbonate (~13 kcal/mole). There is minimal change in the charge on either metal in going from EC to TS1 except for Ph-Co. When TS1 is formed, the charge on the hydroxyl oxygen drops to almost the same values (ranging from −1.02 to −1.09 |eu|) for all eight complexes even though the Zn^{2+} complexes were found to possess higher charges in the EC structures.

After passing the first transition state, a bicarbonate complex directly chelated to the metal is formed. There has been great deal of debate about the actual conformation of the bicarbonate around the metal center in carbonic anhydrase. From these calculations and others, a Lindskog intermediate (OH of bicarbonate is oriented towards the metal, Figure 2) will clearly be formed first in this reaction (denoted I1) [51], [53], [55]. The geometry of both N3 complexes is very similar with one oxygen directly coordinated to the metal center (1.886 Å and 1.900 Å for Co^{2+} and Zn^{2+}, respectively), and the bicarbonate hydroxyl group weakly interacting with the metal ion (2.972 Å and 2.949 Å for Co^{2+} and Zn^{2+}, respectively). The geometry around the metal is tetrahedral. Similar asymmetrical bicarbonate coordination geometries for I1 were obtained for the Ph and Ben complexes for both metal ions (Figure 3). The N4-Zn structure resembles the N3-metal structures with a single oxygen coordinated to Zn^{2+}, and the hydroxyl group asymmetrically interacting with the zinc (1.909 and 2.933 Å). The I1 N4-Co geometry differs from the other complexes. The oxygens coordinating Co^{2+} are much more symmetrical. The metal to coordinating oxygen distance is 1.976 Å, and the hydroxyl oxygen is 2.398 Å away. Although not perfectly octahedral, this structure shows cobalt's ability/preference to coordinate six ligands.

Rotation about the oxygen bond coordinated to the metal center in the Lindskog intermediate (I1) leads through a shallow transition state (TS2) to the lower energy Lipscomb intermediate (I2), which has both carboxylate oxygens of bicarbonate directed towards the metal [56]. This second transition state occurs when the dihedral angle (OC – C, see Figure 1C) has rotated approximately 90°. For both metal ion complexes of N3, Ph, and Ben structures, TS2 has almost identical geometries. Interestingly, the TS2 structures for N4 differ significantly. N4-Zn has a transition state that resembles the N3 structures, but the N4-Co TS2 structure still shows a preference for octahedral binding even though one site is unoccupied. It should be pointed out that proton transfer from the hydroxyl oxygen (OH) of bicarbonate to the non-coordinated oxygen (O) is also a viable mechanism for conversion of I1 to I2 but requires additional water molecules for this to have an activation barrier as low as bond rotation [57]. In either case, this portion of the reaction is not expected to be rate-limiting.

The Lipscomb intermediates for both N3 complexes are similar; a single oxygen is coordinated to the metal, and the other carboxylate oxygen is weakly coordinated to the metal (1.920 and 2.964 Å for Zn^{2+} and 1.905 and 3.027 Å for Co^{2+}). For the N4 complexes, the carboxylates of bicarbonate are also bound differently in the Lipscomb intermediate depending on the metal. For N4-Zn, the oxygen to zinc distances are 1.930 and 2.909 Å which is similar to the values for the unidentate N3-Zn complex. For N4-Co, the oxygen cobalt distances are almost identical at 2.113 and 2.158 Å, again reflecting cobalt's preference for an octahedral geometry in this macrocycle. The bicarbonate geometries for the Ph compounds were unidentate for Zn^{2+} but bidentate for Co^{2+}. These calculated results are in good agreement with crystal structure data of Zn^{2+} and Co^{2+} bound by a tris(pyrazoyl)hydroborato ligand and coordinating nitrate or carbonate [58], [59]. In the Zn^{2+} compounds, only one nitrate or carbonate oxygen binds to the metal at a distance of 1.98 Å, and the second oxygen is greater than 2.6 Å from the metal. For the Co^{2+} compounds, the two oxygens bind more symmetrically around the metal at 2.001 and 2.339 Å for nitrate and 2.055 and 2.271

Å for carbonate in the crystal structures. The bicarbonate I2 geometries for the Ben compounds were almost identical. Both metals bind the bicarbonate in a unidentate geometry with oxygen distances of 1.936 and 3.187 Å for Zn^{2+} and 1.934 and 3.198 Å for Co^{2+}.

The calculated results for nucleophilic attack of CO_2 are in qualitative agreement with model studies. The x-ray crystal geometries of tris(pyrazolyl)hydroborato zinc hydroxide and cobalt hydroxide complexes are similar to those obtained for the N3 complexes [60]. These structures all have tetrahedral geometries around the metal center and readily react with CO_2 to form bicarbonate. Unfortunately, the tris(pyrazolyl)hydroborato complexes are not soluble in water therefore release of the metal bound bicarbonate is not possible with these catalysts.

Product Release

To study the release of bicarbonate from the metal center, a single water molecule was added to the I1 (Lindskog) and I2 (Lipscomb) structures since it was not obvious which geometry would have the lower activation barrier for product release. Once a water molecule was added to each intermediate, the structures were reoptimized. In all cases, the I2 structure with water was the lower energy structure. The water molecule was stabilized by formation of a hydrogen bond between the oxygen of water (OW) and the hydrogen from the amine group in the ring structure of both N3 and N4 and interaction of a hydrogen from water (H1) with the oxygen of bicarbonate (OC) coordinating the metal. For the Ph and Ben structures, the water hydrogen bonds with the bicarbonate but likely does not interact strongly with the rest of the complex since there are no other polar groups in the vicinity. Interestingly, the structure obtained for the Ph-Zn is very similar to the x-ray structures of 2VVB [61] and 1XEG [62], where a unidentate bicarbonate or acetate is interacting with one water molecule. The original intermediate structures were not significantly affected by the inclusion of the water molecule. The energy difference of I1 relative to I2 did not change significantly by adding the water molecule. Unlike the macrocycles, lower energy structures than the encounter complex were found when the water molecule directly coordinates to the metal ion for both Ph and Bencomplexes. When the water coordinates to the metal ion in the Ph complex, a trigonal bipyramidal metal center is formed. Two conformers are possible in the case of zinc, the lowest energy structure has the water in the axial position and the bicarbonate (in the equatorial position) hydrogen bonds to the coordinated water (Figure 5A). Although it is possible to obtain a minimum energy structure with bicarbonate in the axial position and water equatorial (Figure 5B), the energy barrier (TS_{turn}) separating these two structures disappears when the ZPE correction is included. A turnstile pseudorotation occurs to transform one structure to the other, but the amount of rearrangement to have the bicarbonate go from axial to equatorial is small because of the three-fold symmetry of the phosphine complex. Both ligands only need to rotate by ~60° to interconvert between conformations. Similar coordination of water and bicarbonate around the zinc has also been observed in carbonic anhydrase II binding acetate.

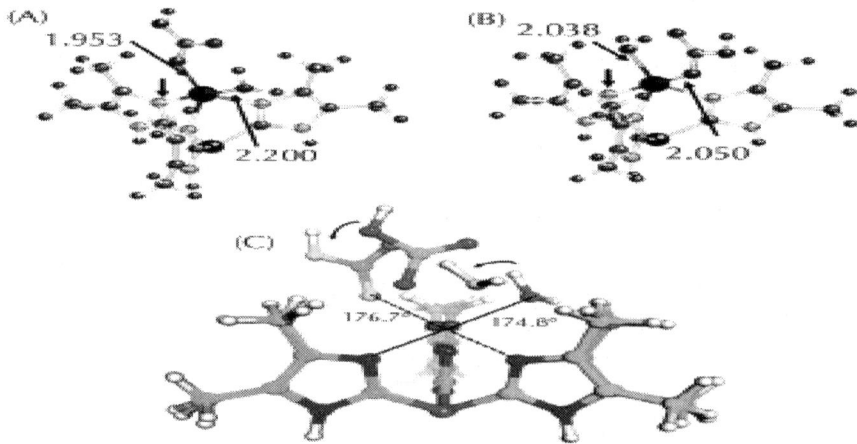

Figure 5: Optimized structures of the Ph-Zn-bicarbonate-water complexes. Panel (A) is the lowest energy structure and has the water in the axial position. The angle formed by the imidazole nitrogen (arrow)-Zn-oxygen (water) is almost linear. Panel (B) has the bicarbonate in the axial position. Panel (C) shows an overlay of the two structures and how interconversion between the two geometries can occur.

In the wild-type (WT) protein (PDB, 1CAY), the coordination around the zinc was a distorted trigonal bipyrimid with a water as an equatorial ligand and the acetate as an axial ligand [63]. Comparison of the WT enzyme with the T199A mutant (PDB, 1CAM) shows that the hydroxyl group of T199 is important in the positioning of the water molecule coordinated to the zinc [64]. The positioning of the water and the bicarbonate around the zinc by Thr199 likely creates a situation in which the geometry is not optimal and makes release of bicarbonate more favorable. The Zn-carboxylate oxygen distances for the WT and T199A proteins are 2.42 and 2.27 Å, respectively. In the 1CAM crystal structure, the angle formed by the His94 NE2– Zn – O of water is 136.6°. The analogous angle in 1CAY is 110.0°. Experimentally, the T199A mutant is ~100 times slower at turning over CO_2 than the WT enzyme, and the binding of inhibitors such as thiocyanate and bicarbonate is enhanced by 20-fold [65]. In the E106Q mutant protein (PDB, 1CAZ) [63], the zinc coordination is trigonal bipyramidal, but the water and acetate coordination is now reversed with water as the axial ligand and acetate as the equatorial ligand. These calculations show bicarbonate is more strongly coordinated to the Zn^{2+} in the equatorial position (Zn-O bond is 1.953 Å) relative to the axial position (Zn-O bond is 2.050 Å) in Ph-Zn and the lowest energy conformation for the trigonal bipyramidial geometry. The carboxylate sidechain of E106 is important for positioning the water around the zinc to avoid this conformation since the E106D mutant shows little change in activity from the wild-type enzyme. The amide sidechain of E106Q rotates away from T199 and functions as a hydrogen donor with T199 instead of an acceptor. This changes the hydrogen bonding network within the active site, resulting in a 1000-fold decrease in the maximal rate for the E106Q mutant [65].

The Ph-Co encounter complex is a distorted trigonal bipyramidal structure with water in the axial position and bicarbonate in the equatorial position. One oxygen of the bicarbonate is hydrogen bonding with the water molecule. This complex could also be described as having an octahedral geometry but missing the sixth ligand. Interestingly, a formal octahedral complex 3.6 kcal/mole higher in energy relative to the pentacoordinate structure was also obtained. This structure, which has two oxygens of the bicarbonate equatorially coordinated to Co^{2+} and water at the axial position, is almost identical to the Co^{2+} carbonic anhydrase binding bicarbonate (PDB 1CAH) [66] (Figure 6) and also the structure of cobalt tris[2-isopropylimidazol-4(5)-yl]phosphanecoordinating to nitrate and water [67]. A similar octahedral structure was also obtained for the Zn^{2+} complex that was 5.44 kcal/mole higher in energy relative to the I3 structure. The trigonal bipyramidal coordination of the bicarbonate in the enzyme may not be favorable. An overlay of the I3 structure in the active site of 1CAH shows the bicarbonate would be in close contact with the side chain of L198.

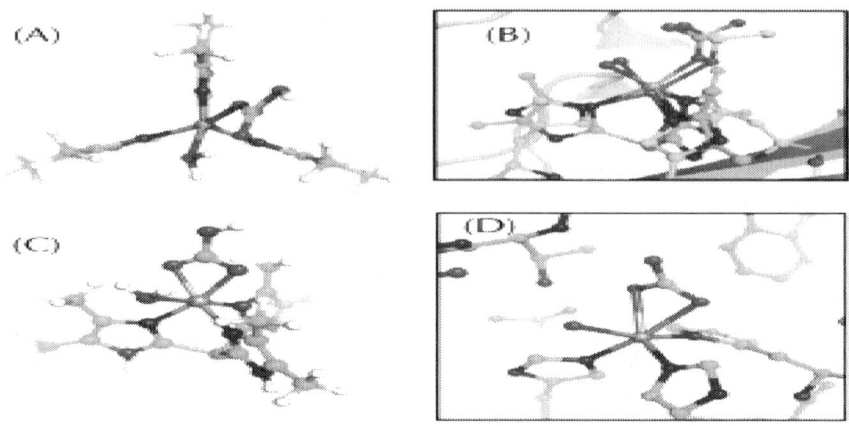

Figure 6: Optimized structures of Ph-Co-bicarbonate-water complexes. Panel (A) shows the low energy structure of Ph-Co interacting with water (I3). Panel (B) shows an overlay of the Ph-Co (I3, green) structure with the wild-type (zinc) carbonic anhydrase with acetate (1XEG, cyan). Panel (C) shows an octahedral geometry for bicarbonate, and water bound to cobalt. Panel (D) shows the arrangement of ligands (water and bicarbonate) around the metal ion in the X-ray crystal structure of cobalt carbonic anhydrase (1CAH).

Axial and equatorial arrangements for bicarbonate in both Ben-metal complexes were also found, with the equatorial geometry the more stable by ~2 kcal/mole, but interconversion between these structures was not possible (Figure 7). When water coordinates to the metal, an octahedral complex forms for both Ben-metal complexes as the bicarbonate shift positions to take up an equatorial arrangement around the metal. The interconversion of conformation by the turnstile pseudorotation in this case would likely have a high barrier since this complex is not symmetric and would require the water molecule and bicarbonate to exchange positions (180° rotation). The octahedral geometry adopted by the Ben-Zn complex is reminiscent of tris (6-amino-2-pyridylmethyl)amine binding Zn^{2+} which catalyzes phosphodiester cleavage [68]. Indeed, complexes of Zn^{2+}, Co^{2+}, and Cu^{2+} coordinated to tris (2-benimidazylmethyl)amine have been shown to catalyze the hydrolysis of p-nitrophenyl acetate [69].

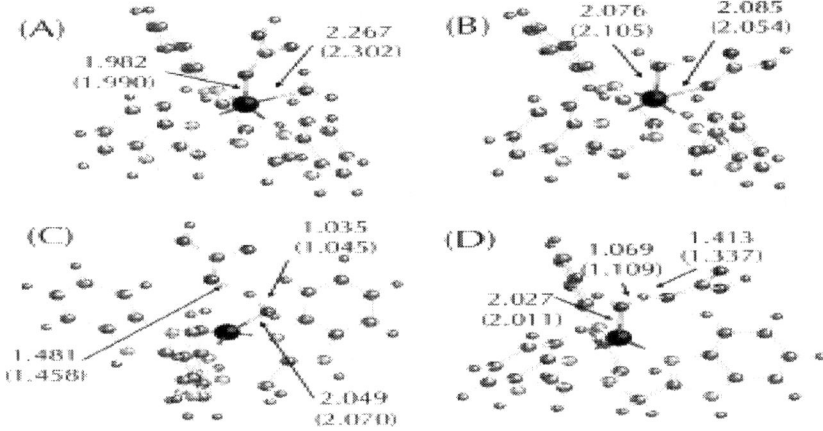

Figure 7: Optimized geometries for the bicarbonate and water bound to Ben-Zn with the bicarbonate in the equatorial position (A) and axial position (B). The equatorial structure is the most stable. Panel (C) and (D) show the corresponding transition states for (A) and (B), respectively. Values are in angstroms and values in parenthesis are for the corresponding cobalt structures.

For product release, an addition-substitution reaction occurs with a water molecule displacing the bicarbonate. At the transition state (TS3), the water molecule coordinates to the metal ion, causing the oxygen of the bicarbonate to weaken (Figure 7 and 8). In the zinc complexes, the oxygen of water is 1.900 Å (N3), 2.024 Å (N4), 1.961 Å (Ph), and 2.049 Å (Ben) from Zn^{2+} with the oxygen of the previously coordinated bicarbonate now at 2.938 Å (N3), 2.886 Å (N4), 2.676 Å (Ph), and 2.823 Å (Ben). The displaced oxygen of bicarbonate interacts with one of the hydrogens of the water and ultimately abstracts the proton from the water to form carbonic acid and to reform the metal-hydroxide catalyst. The transition state geometry for N3-Co is almost identical to N3-Zn, with the water bound slightly tighter (1.961 Å) and the bicarbonate more weakly bound to the metal (3.017 Å). The transition state for N4-Co differs from the other three structures. The water is bound tightly to the Co^{2+}, and the oxygen of bicarbonate is still coordinated to the metal (2.509 Å). Additionally, the previously mentioned TS3 structures (bothN3 and N4-Zn) had the oxygen of bicarbonate interacting with one of the hydrogens on the water molecule. In the TS3 structure of N4-Co, the hydrogen/proton from the water has transferred to the bicarbonate to form carbonic acid.

No change in the ring structure occurred for either N4 complex at the transition state.

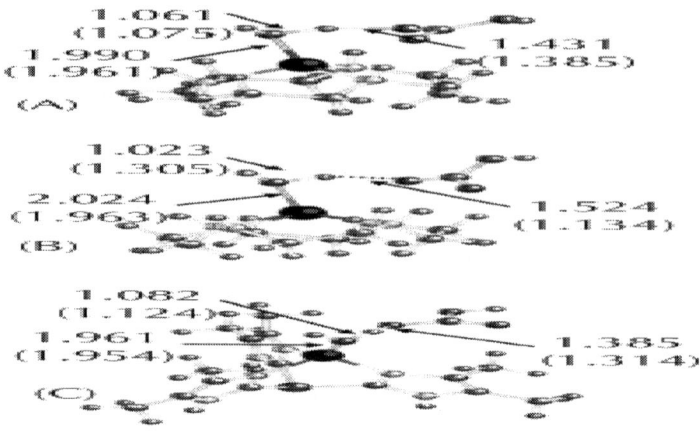

Figure 8: Calculated transition state structures (TS3) for displacement of bicarbonate by a water molecule for N3-Zn (A), N4-Zn (B), and Ph-Zn (C). Numerical values are in angstroms and values in parenthesis are the corresponding values for the cobalt structures.

From these calculations, Ben-Zn should be a poorer catalyst at CO_2 hydration that N3 or N4, although the water soluble sulfonated version of Ben-Zn was reported to be highly active at low temperatures, and its activity was extrapolated to room temperature [70]. Interestingly, no direct kinetic measurements for the sulfonated-Ben-Zn were reported at room temperature. To better understand the catalytic properties of Ben-Zn, the sulfonated benzimidazole compound was synthesized and tested. The sulfonated-Ben-Zn complex did not show any catalytic properties at room temperature and had slight activity at 50°C, which was consistent with the calculated activation barrier for Ben-Zn.

For all complexes, the transition state from the I2 structure was lower in energy than the transition state from the I1 structure. Interestingly, the activation barrier from I1 or I2 to their respective transition states was almost identical in value. The activation barrier for bicarbonate release ranges from a low value of 14.8 kcal/mole for N4-Zn to a high value of 20.7 kcal/mole for Ben-Co. From these calculations, product release is the rate-limiting step for the hydration of CO_2. These values differ from those obtained by Mauksch et al., using the model system [(NH$_3$)$_3$Zn(OH)]$^+$/CO$_2$ [71]. They find that nucleophilic attack is the rate-limiting step for CO$_2$ hydration and only a small barrier for product release. This discrepancy is due to their assumption that one of the protons on the coordinated water molecule transfers to the bicarbonate while both are still coordinated to the zinc. This proton transfer seems unlikely in solution from pK_a measurement of the macrocycle triamine [2-(2-hydroxyphenyl)-1,5,9-triazacyclodo decanecoordinated with zinc [72]. This macrocycle is pentacoordinated with a trigonal bipyramidal geometry. The 2-hydroxyphenyl moiety has a pK_a of 6.8, and the coordinated water has a pK_a of 10.7. Having a charged oxygen coordinated to the zinc reduces the metal's ability to acidify the water molecule since the pK_a of water bound to 1,5,9-triazacyclododecane-zinc is 7.5. The calculated activation barriers for the zinc complexes forN4 (14.8 kcal/mole), N3 (15.7 kcal/mole), Ph (15.6 kcal/mole), and Ben (18.3 kcal/mole) are in reasonable agreement with measured rate constants for CO$_2$ hydration 2494 M^{-1} S^{-1} [73], 1083 M^{-1} S^{-1} [73], 898 M^{-1} S^{-1} [74], and not catalytic, respectively. The correlation between product release and experimental rate constants is consistent with our previous results, showing the bond dissociation energy between bicarbonate and Zn-azamacrocycles corresponds with the experimental results [73].

The calculated hydration reaction catalyzed by the tetrahedral coordinating N3, using either Zn^{2+} or Co^{2+}, was very similar in both geometries and energies obtained. This is consistent with experimental results that show almost identical coordination geometries and wild-type activity for alpha-class carbonic anhydrases that have Zn^{2+} substituted with Co^{2+} [17], [75]. The calculated activation barrier for release of bicarbonate is high in these polyamine complexes yet HCAII experiments have shown this step in the reaction to be rapid and not rate-limiting [76]. In HCAII, both experiment and theory have shown that Thr199 has a destabilizing effect on bicarbonate binding to zinc [65], [77]. Hybrid QM/MM calculations by Merz and Banci show that the active site of HCAII promotes destabilization by pulling one of the carboxylate oxygens of bicarbonate away from the zinc by formation of a hydrogen bond with the hydroxyl group of Thr199 [78]. Using PM3 calculations, they found that having the zinc-bicarbonate active site geometry destabilizes the Lipscomb intermediate by 8.7 kcal/mole relative to the QM optimized structure. These results are also qualitatively in agreement with estimated free energies from kinetics data for carbonic anhydrase that show dissociation of bicarbonate limits the CO_2 hydration catalyzed by HCAI and the Thr200His mutant of HCAII [79].

Solvent Effects on CO_2 Hydration

To estimate the effects of solvent on the CO_2 hydration reaction, single-point conductor-like polarization continuum model (CPCM) calculations were performed on the optimized gas-phase geometries. Addition of solvation effects removes the encounter complex as a minimum along the reaction coordinate (Figure 9). The separated reactants go directly to the first transition state, and the activation barrier is significantly lowered. The activation barrier ranged from 0.21 to 3.16 kcal/mole. Once past the transition state, the bicarbonate is formed. When including solvation effects, the energy differences between the Lindskog (I1) and Lipscomb (I2) intermediates are much smaller. In the gas-phase the energy difference was ~5 kcal/mole or greater, but in solution the energy differences are reduced and ranged from 0.18 to 2.31 kcal/mole. Addition of a water molecule to the intermediate structures does not significantly change the energy difference between the two geometries for the CPCM calculations. In some cases, the activation barrier for interconversion of I1 to I2 is the highest barrier.

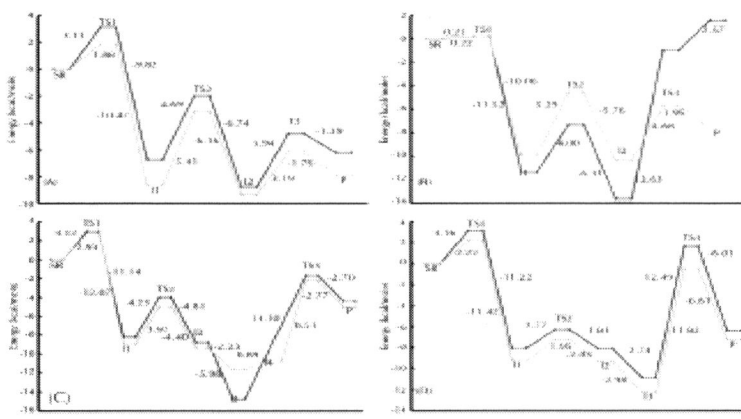

Figure 9: Relative energies to the separated reactants from single point solvation (CPCM) calculations using the gas-phase stationary points for the zinc

(black) and cobalt (gray) complexes are shown for N3 (A), N4 (B), Ph (C), and Ben (D).

The activation barrier for bicarbonate release is significantly reduced by solvent effects. In the case of the N3 complexes, the activation barrier is reduced by 12.48 and 11.56 kcal/mole for N3-Zn and N3-Co, respectively. Solvation in the high dielectric medium makes separation of the two charged species more favorable, and the reduction in the barrier is large enough that in the case of N3-metal ion, interconversion of I1 to I2 is rate-limiting. Having the activation barrier for interconversion between the two intermediate structures as the rate-limiting step for this reaction is an unexpected result, and may be an artifact of using the gas-phase geometries. This is the only transition state (TS2) along the reaction coordinate that does not involve either formation or breaking a bond. TS1 and TS3 maybe over stabilized when using the gas-phase geometries. A similar result occurred in a computational study of CO_2 being converted to HCO_3^- by a -carbonic anhydrase containing either cadmium or zinc [52]. In the gas-phase, nucleophilic attack of CO_2 had the highest activation barrier, but when the dielectric constant was set to 4 (to better simulate the protein environment), the rotation barrier was rate-limiting. Interestingly, when the dielectric constant was set to 80, the rate-limiting step was again the nucleophilic attack. It should also be noted that we chose to model the interconversion of the intermediates by rotation around the bicarbonate for computational ease. Since the bicarbonate is solvent exposed on these mimetics, proton exchange with water might be a lower energy route for interconversion.

With the inclusion of solvation, the activation barrier for bicarbonate release is almost identical for either the Lindskog or Lipscomb intermediates for both N3 complexes and N4-Zn. The active site of HCAII has a well ordered solvent network that provides a route for proton release to bulk solvent [80], but this network may also contribute in lowering the barrier for product release. The CPCM activation barriers for product release from Zn^{2+} overestimate the stabilization for the calculated barrier for bicarbonate dissociation in the macrocycles. Loferer et al., using QM/MM methods [81], calculated a barrier of 6.2 kcal/mole for bicarbonate dissociation in CAII, and from these calculations barriers of 3.19 (N3-Zn) and 4.68 kcal/mole (N4-Zn) were obtained. Interestingly, the activation barriers for product release are much higher for Ph-Zn (9.40 kcal/mole) and Ben-Zn (11.82 kcal/mole). Having a hydrophobic environment around the reacting species (methyl groups in Ph and phenyl groups in Ben) shows that nucleophilic attack is sensitive to solvent exposure, but the product release barrier is less affected by solvent. In fact, other than nucleophilic attack of CO_2, the reaction profile for both Ph and Ben are relatively unchanged. In the gas-phase, the overall reaction for all species is endothermic, but with the addition of solvation effects all the reactions become exothermic with the exception of N4-Co.

For N4-Co, solvation effects did not significantly reduce the activation barrier for bicarbonate release and is the only species that is endothermic in solvent. It would appear that N4-Co does not catalyze the hydration of CO_2 even though it had the lowest barrier for nucleophilic attack. The preference of an octahedral geometry for N4-Co makes release of bicarbonate improbable. This is consistent with experiments that showed a 5-coordinated Co^{2+} complex (four nitrogens and one oxygen from water) is able to form cobalt-hydroxide but does not catalyze the hydration of CO_2 [82], [83]. We should also point out that having an additional water molecule coordinated to the cobalt could contribute to lowering the activation barrier or change the coordination of bicarbonate to unidentate, but we did not pursue these calculations since it was beyond the scope of the present study. Clearly, solvation has a significant effect on the activation barrier for product release, although the reduction in the barrier could be overestimated because we are not using optimized CPCM structures.

CONCLUSIONS

Models that mimic the reactivity of carbonic anhydrase are of interest not only academically but to industry, which is trying to lower the amount of CO_2 being released into the atmosphere [84]. Two of the most successful mimetics of carbonic anhydrase are the cyclic polyamines, 1,4,7,10-tetraazacyclododedacane (N4) and 1,5,9-triazacyclododedacane (N3), and when coordinated to zinc are able to catalyze the reversible hydration of CO_2. From our calculations, the Zn^{2+} and Co^{2+} complexes of N3 have very similar coordination geometries to human carbonic anhydrase II and comparable energetics. The N4-Zn complex has slightly higher turnover than the N3-Zn but has been criticized as a mimic for human carbonic anhydrase II because it has pentacoordinate geometry. Although the coordination differs, the calculations show that N4-Zn follows the same reaction as the N3-Zn/Co complexes. The N4-Co complex is able to lower the barrier for nucleophilic attack more than any of the other complexes by having an octahedral geometry around Co^{2+}, but this is at the expense of being able to release bicarbonate later in the reaction.

Interestingly, the gamma-class carbonic anhydrase from *Methanosarcina thermophila* (Cam), which normally uses Fe^{2+} to catalyze the hydration of CO_2, may have found a way around this product release problem [85]. This carbonic anhydrase can also utilize pentacoordinated Zn^{2+} or hexacoordinated Co^{2+}, and Co-Cam is actually better at catalyzing the hydration reaction than Zn-Cam [86]. The crystal structure of bicarbonate bound in Zn-Cam and Co-Cam show they have different coordination positions around the metal ion [87]. For Zn-Cam, the geometry of bicarbonate resembles the Lipscomb intermediate for N3 with the carboxylate oxygens 2.48 and 3.11 Å from Zn^{2+}. In Co-Cam, only one oxygen in bicarbonate is bound to Co^{2+}, and two water molecules take up the other coordination sites. Interestingly, the geometry of bicarbonate around Co^{2+} most resembles the TS2 structure of N4-Co. It would be interesting if a catalyst based on the binding geometry in Cam could be created. If possible, its application to industry could be significant since the susceptibility of Zn-Cam and Co-Cam to anionic inhibitors differs [88]. The difference in the inhibitors is likely due to the coordination preference of the metals.

The activation barriers for N3-Zn and N4-Zn from our calculations are quite low yet these complexes are ~1000 slower at catalyzing the hydration of CO_2 relative to HCAII. One aspect of the reaction that could not be readily studied is the importance of reactant positioning. Although the rate-limiting step in HCAII is proton loss from the metal bound water, it would not be expected to be limiting for these mimetics that are solvent exposed and function optimally at alkaline pH. Recent crystal structures show that HCAII contains a hydrophobic pocket that binds CO_2 in a conformation that will readily react with the zinc-hydroxide [54]. In fact, the presence of a metal is not even required for CO_2 binding in HCAII. Additionally, placement of zinc coordinated by three histidines within the hydrophobic interior of an -helical triple coiled coil showed CO_2 hydration activity within 500-fold of HCAII

[89]. Reactant positioning likely is an important aspect of the hydration reaction by HCAII and for these mimetics.

ACKNOWLEDGMENTS

This work performed under the auspices of the U.S. Department of Energy by Lawrence Livermore National Laboratory under Contract DE-AC52-07NA27344. A generous allocation of computer time on Livermore Computing is gratefully acknowledged. LLNL-JRNL-589234.

AUTHOR CONTRIBUTIONS

Conceived and designed the experiments: JHS RDA FCL. Performed the experiments: EYL SEW LK CAV JPB SEB. Analyzed the data: JHS RDA FCL EYL. Contributed reagents/materials/analysis tools: CAV. Wrote the paper: EYL SEW RDA SEB CAV.

REFERENCES

1. Pachauri R, Reisinger A (2007) "Climate Change 2007: Synthesis Report," Intergovernmental Panel on Climate Change.
2. Figueroa J, Fout T, Plasynski S, McIvried H, Srivastava R (2008) Advances in CO_2 capture technology–The U.S. Department of Energy's Carbon Sequestration Program. Int. J. Greenhouse Gas Control 2: 9–20. doi: 10.1016/s1750-5836(07)00094-1
3. Stumm W, Morgan J (1996) Aquatic Chemistry; 3rd ed.; Wiley-Interscience.
4. Stolaroff JK, Keith DW, Lowry GV (2008) Carbon Dioxide Capture from Atmospheric Air Using Sodium Hydroxide Spray, Environ. Sci. Technol. 4: 2728–2735. doi: 10.1021/es702607w
5. Stolaroff JK (2006) Capturing CO_2 from ambient air: A feasibility assessment, PhD thesis, Carnegie Mellon University.
6. Cullinane JT, Rochelle GT (2006) Kinetics of Carbon Dioxide Absorption into Aqueous Potassium Carbonate and Piperzine. Ind. Eng. Chem. Res. 45: 2531–2545. doi: 10.1021/ie050230s

7. Bao LH, Trachtenberg MC (2006) Facilitated transport of CO2 across a liquid membrane: Comparing enzyme, amine, and alkaline. J. Membrane Sci. 280: 330–334. doi: 10.1016/j.memsci.2006.01.036
8. Elliott S, Lackner KS, Ziock HJ, Dubey MK, Hanson HP, et al. (2001) Compensation of atmospheric CO2 buildup through engineered chemical sinkage. Geophys. Res. Lett. 28: 1235–1238. doi: 10.1029/2000gl011572
9. Keith DW, Ha-Duong M, Stolaroff JK (2005) Climate Strategy with CO2 Capture from the Air. Climatic Change 74: 17–45. doi: 10.1007/s10584-005-9026-x
10. Lindskog S (1997) Structure and mechanism of carbonic anhydrase. Pharmacol. Ther. 74: 1–20. doi: 10.1016/s0163-7258(96)00198-2
11. Steiner H, Jonsson BH, Lindskog S (1975) The catalytic mechanism of carbonic anhydrase. Hydrogen-isotope effects on the kinetic parameters of the human C isoenzyme. Eur. J. Biochem. 59: 253–259. doi: 10.1111/j.1432-1033.1975.tb02449.x
12. Silverman DN, Lindskog S (1998) The catalytic mechanism of carbonic anhydrase: implications on the rate-limiting protolysis of water. Acc. Chem. Res. 21: 30–36. doi: 10.1021/ar00145a005
13. Khalifah RG (1971) The Carbon Dioxide Hydration Activity of Carbonic Anhydrase I. Stop-flow kinetic studies on the native human isoenzymes B and C. J. Biol. Chem. 246: 2561–2573.
14. Liljas A, Kannan KK, Bergsten PC, Waara I, Fridborg K, et al. (1972) Crystal Structure of Human Carbonic Anhydrase C. Nat. New Biol. 235: 131–137. doi: 10.1038/newbio235131a0
15. Hakansson K, Carlsson M, Svensson LA, Liljas A (1992) Structure of the native and apo carbonic anhydrase II and structure of some of its anion-ligand complexes. J. Mol. Biol. 227: 1192–1204. doi: 10.1016/0022-2836(92)90531-n
16. Lindskog S, Nyman PO (1964) Metal-binding properties of human erythrocyte carbonic anhydrase. Biochim. Bipohys. Acta 85: 462–474. doi: 10.1016/0926-6569(64)90310-4
17. Kogut KA, Rowlett RS (1987) A comparison of the mechanisms of CO2 hydration by native and Co2+-substituted carbonic anhydrase II. J. Biol. Chem. 262: 16417–16424.

18. Bertini I, Luchinat C (1983) Cobalt (II) as a probe of the structure and function of carbonic anhydrase. Acc. Chem. Res. 16: 272–279. doi: 10.1021/ar00092a002
19. Cowan RM, Ge JJ, Qin YJ, McGregor ML, Trachtenberg MC (2003) CO2 by means of an enzyme-based reactor. Ann. NY. Acad. Sci. 984: 453–469. doi: 10.1111/j.1749-6632.2003.tb06019.x
20. Bhattacharya S, Nayak A, Schiavone M, Bhattacharya SK (2004) Solubilization and concentration of carbon dioxide: novel spray reactors with immobilized carbonic anhydrase. Biotechnol. Bioeng. 86: 37–46. doi: 10.1002/bit.20042
21. Azari F, Nemat-Gorgani M (1999) Reversible denaturation of carbonic anhydrase provides a method for its adsorptive immobilization. Biotechnol. Bioeng. 62: 192–199. doi: 10.1002/(sici)1097-0290(19990120)62:2<193::aid-bit9>3.0.co;2-h
22. Yan M, Liu ZX, Lu DN, Liu Z (2007) Fabrication of single carbonic anhydrase nanogel against denaturation and aggregation at high temperatures. Biomacromolecules 8: 560–565. doi: 10.1021/bm060746a
23. Krishnamurthy VM, Kaufman GK, Urbach AR, Gitlin I, Gudiksen KL, et al. (2008) Carbonic anhydrase as a model for biophysical and physical-organic studies of proteins and protein-ligand binding. Chem Rev. 108: 946–1051. doi: 10.1021/cr050262p
24. Parkin G (2004) Synthetic Analogues Relevant to the Structure and Function of Zinc Enzymes. Chem. Rev. 104: 699–767. doi: 10.1021/cr0206263
25. Zhang XP, van Eldik R, Koike T, Kimura E (1993) Kinetics and Mechanism of the Hydration of Carbon Dioxide and Dehydration of Bicarbonate Catalyzed by a Zinc(II) Complex of 1,5,9-Triazacyclododecane as a Model for Carbonic Anhydrase. Inorg. Chem. 32: 5749–5755. doi: 10.1021/ic00077a017
26. Zhang XP, van Eldik R (1995) A Functional Model for Carbonic Anhydrase: Thermodynamics and Kinetics Study of a Tetraazacycledodecane Complex of Zinc(II). Inorg. Chem. 34: 5606–5614. doi: 10.1021/ic00126a034
27. Frisch MJ, Trucks GW, Schlegel HB, Scuseria GE, Robb MA, et al. Gaussian 03, Revision C.02, Gaussian, Inc., Wallingford CT, 2004.

28. Frisch MJ, Trucks GW, Schlegel HB, Scuseria GE, Robb MA, et al. Gaussian 09, Revision B.01 Gaussian, Inc., Wallingford CT, 2009.
29. Becke AD (1993) Density-functional thermochemistry. III. The role of exact exchange. J. Chem. Phys. 98: 5648–5652. doi: 10.1063/1.464913
30. Lee C, Yang W, Parr RG (1988) Development of the Colle-Salvetti correlation-energy formula into a functional of the electron density. Phys. Rev. B 37: 785–789. doi: 10.1103/physrevb.37.785
31. Haffner PE, Coleman JE (1973) High Spin and Low Spin Forms of Co(II) Carbonic Anhydrase: Temperature-Dependent Changes in Spin State and Coordination Geometry. J. Biol. Chem. 248: 6630–6636.
32. Boys SF, Bernardi F (1970) The calculation of small molecular interactions by the difference of separate total energies. Some procedures with reduced errors. Mol. Phys. 19: 553–566. doi: 10.1080/00268977000101561
33. Schultz NE, Zhao Y, Truhlar DG (2005) Density functionals for inorganometallic and organometallic chemistry. J. Phys. Chem. A 109: 11127–11143. doi: 10.1021/jp0539223
34. Barone V, Cossi M (1998) Quantum Calculation of Molecular Energies and Energy Gradients in Solution by a Conductor Solvent Model. J. Phys. Chem. A. 102: 1995–2001. doi: 10.1021/jp9716997
35. Barone V, Cossi M, Tomasi J (1998) Geometry optimization of molecular structures in solution by the polarizable continuum model. J. Comput. Chem. 19: 404–417. doi: 10.1002/(sici)1096-987x(199803)19:4<404::aid-jcc3>3.0.co;2-w
36. Mineva T, Russo N, Sicilia E (1998) Solvation effects on reaction profiles by the polarizable continuum model coupled with the Gaussian density functional method. J. Comput. Chem. 19: 290–299. doi: 10.1002/(sici)1096-987x(199802)19:3<290::aid-jcc3>3.0.co;2-o
37. Barone V, Cossi M, Tomasi J (1997) A new definition of cavities for the computation of solvent free energies by the polarizable continuum model. J. Chem. Phys., 1997 107: 3210–3221. doi: 10.1063/1.474671

38. Rappe AK, Casewit CJ, Colwell KS, Goddard III WA, Skiff WM (1992) UFF, a Full Periodic Table Force Field for Molecular Mechanics and Molecular Dynamics Simulations. J. Am. Chem. Soc. 114: 10024–10035. doi: 10.1021/ja00051a040
39. Reed AE, Weinstock RB, Weinhold F (1985) Natural population analysis. J. Chem. Phys. 83: 735–746. doi: 10.1063/1.449486
40. da Silva Miranda F, Signori AM, Vicente J, de Souza B, Priebe JP, et al. (2008) Synthesis of substituted dipyrido[3,2-a:2',3'-c]phenazines and a new heterocyclic dipyrido[3,2-f:2',3'-h]quinxoalino[2,3-b]quinoxaline. Tetrahedron 64: 5410–5415. doi: 10.1016/j.tet.2008.02.097
41. Rodinov VO, Presolski SI, Gardinier S, Lim YH, Finn MG (2007) Benzimidazole and Related Ligands for Cu-Catalyzed Azide-Alkyne Cycloaddition. J. Am. Chem. Soc. 129: 12696–12704. doi: 10.1021/ja072678l
42. Bertini I, Lanini G, Luchinat C (1983) Equilibrium species in cobalt(II) carbonic anhydrase. J. Am. Chem. Soc. 105: 5116–5118. doi: 10.1021/ja00353a043
43. Yachandra V, Powers L, Spiros TG (1983) X-ray Absorption Studies and the Coordination Number of Zn and Co Carbonic Anhydrase as a Function of pH and Inhibitor Binding. J. Am. Chem. Soc. 105: 6596–6604. doi: 10.1021/ja00360a009
44. Sola M, Lledos A, Duran M, Bertran J (1992) Ab initio study of the hydration of carbon dioxide by carbonic anhydrase. A comparison between the Lipscomb and Lindskog mechanisms. J. Am. Chem. Soc. 114: 869–877. doi: 10.1021/ja00029a010
45. Zheng YJ, Merz KM (1992) Mechanism of the human carbonic anhydrase II-catalyzed hydration of carbon dioxide. J. Am. Chem. Soc. 114: 10498–10507. doi: 10.1021/ja00052a054
46. Aqvist J, Fothergill M, Warshel A (1993) Computer Simulation of the CO2/HCO3- Interconversion Step in Human Carbonic Anhydrase I. J. Am. Chem. Soc. 115: 631–635. doi: 10.1021/ja00055a036
47. Lu D, Voth GA (1998) Proton Transfer in the Enzyme Carbonic Anhydrase: An ab initio Study. J. Am. Chem. Soc. 120: 4006–4014. doi: 10.1021/ja973397o
48. Bottoni A, Lanza CZ, Miscione GP, Spinelli D (2004) New model for a theoretical density functional theory investigation of the

mechanism of the carbonic anhydrase: how does the internal bicarbonate rearrangement occur? J. Am. Chem. Soc. 126: 1542–1550. doi: 10.1021/ja030336j

49. Miscione GP, Stenta M, Spinelli D, Anders E, Bottoni A (2007) New computational evidence for the catalytic mechanism of carbonic anhydrase. Theor. Chem. Acc., 2007 118: 193–201. doi: 10.1007/s00214-007-0274-x

50. Garmer DR, Krauss M (1992) Metal substitution and the active site of carbonic anhydrase. J. Am. Chem. Soc. 114: 6487–6493. doi: 10.1021/ja00042a031

51. Marino T, Russo N, Toscano M (2005) A comparative study of the catalytic mechanisms of the zinc and cadmium containing carbonic anhydrase. J. Am. Chem. Soc. 127: 4242–4253. doi: 10.1021/ja045546q

52. Amata O, Marino T, Russo N, Toscano M (2011) Catalytic activity of a -class zinc and cadmium containing carbonic anhydrase. Compared work mechanisms. Phys. Chem. Chem. Phys. 13: 3468–3477. doi: 10.1039/c0cp01053g

53. Brauer M, Perez-Lustres JL, Weston J, Anders E (2002) Quantitative reactivity model for the hydration of carbon dioxide by biomimetic zinc complexes. Inorg. Chem. 41: 1454–1463. doi: 10.1021/ic0010510

54. Domsic JF, Avvaru BS, Kim CU, Gruner SM, Agbandje-McKenna M, et al. (2008) Entrapment of carbon dioxide in the active site of carbonic anhydrase II. J. Biol. Chem. 283: 30766–30771. doi: 10.1074/jbc.m805353200

55. Merz KM, Hoffmann R, Dewar MJS (1989) The mode of action of carbonic anhydrase. J. Am. Chem. Soc. 111: 5636–5649. doi: 10.1021/ja00197a021

56. Liang JY, Lipscomb WN (1987) Hydration of carbon dioxide by carbonic anhydrases: internal proton transfer of Zn^{2+}-bound HCO_3^-. Biochemistry 26: 5293–5301. doi: 10.1021/bi00391a012

57. Tautermann CS, Loferer MJ, Voegele AF, Liedl KR (2003) About the Kinetic Feasibility of the Lipscomb Mechanism in Human Carbonic Anhydrase II. J. Phys. Chem. B 107: 12013–12020. doi: 10.1021/jp0353789

58. Han R, Looney A, McNeill K, Parkin G, Rheingold AL, et al. (1993) Structural and spectroscopic studies on four-, five-, and six-coordinate complexes of zinc, copper, nickel, and cobalt: Structural models for the bicarbonate intermediate of the carbonic anhydrase cycle. J. Inorg. Biochem. 49: 105–121. doi: 10.1016/0162-0134(93)85020-9

59. Kitajima N, Hikichi S, Tanaka M, Moro-oka T (1993) Fixation of Atmospheric CO_2 by a Series of Hydroxo Complexes of Divalent Metal Ions and the Implication for the Catalytic Role of Metal Ion in Carbonic Anhydrase. Synthesis, Characterization, and Molecular Structure of [LM(OH)]n (n = 1 or 2) and LM(μ-CO_3) ML (M(II) = Mn, Fe, Co, Ni, Cu, Zn; L = HB(3,5-iPr2pz)3). J. Am. Chem. Soc. 115: 5496–5508. doi: 10.1021/ja00066a018

60. Bergquist C, Fillebeen T, Morlok MM, Parkin G (2003) Protonation and Reactivity towards Carbon Dioxide of the Mononuclear Tetrahedral Zinc and Cobalt Hydroxide Complexes, [TpBut,Me] ZnOH and [TpBut,Me]CoOH: Comparison of the Reactivity of the Metal Hydroxide Function in Synthetic Analogues of Carbonic Anhydrase. J. Am. Chem. Soc. 125: 6189–6199. doi: 10.1021/ja034711j

61. Sjoeblom B, Polentarutti M, Djinovic-Carugo K (2009) Structural study of the X-ray induced activation of carbonic anhydrase. Proc. Natl. Acad. Sci. USA. 106: 10609–10613. doi: 10.1073/pnas.0904184106

62. Mazumdar PA, Kumaran D, Das AK, Swaminathan S (2008) A novel acetate-bound complex of human carbonic anhydrase II, Acta. Crystallogr. Sect. F 64: 163–166. doi: 10.1107/s1744309108002078

63. Hakansson K, Briand C, Zaitsev V, Xue Y, Liljas A (1994) Wild-type and E106Q mutant carbonic anhydrase complexed with acetate. Acta, Crystallogr. Sect. D. 50: 101–104. doi: 10.1107/s0907444993009667

64. Xue Y, Liljas A, Jonsson BH, Lindskog S (1993) Structural analysis of the zinc hydroxide-Thr-199-Glu-106 hydrogen-bond network in human carbonic anhydrase II. Proteins, Struc., Funct., and Genetics. 17: 93–106. doi: 10.1002/prot.340170112

65. Liang Z, Xue Y, Behravan G, Jonsson BH, Lindskog S (1993) Importance of the conserved active-site residues Tyr7, Glu106 and Thr199 for the catalytic function of human carbonic anhydrase II.

Eur. J. Biochem. 211: 821–827. doi: 10.1111/j.1432-1033.1993.tb17614.x

66. Hakansson K, Wehnert A (1992) Structure of cobalt carbonic anhydrase complexed with bicarbonate. J. Mol. Biol. 228: 1212–1218. doi: 10.1016/0022-2836(92)90327-g

67. Kunz PC, Zribi A, Frank W, Klaui W (2007) Synthesis and Characterization of Water-Soluble Zinc, Cobalt(II) and Copper(II) Complexes with a Neutral Tripodal N,N,N-Ligand: Crystal Structures of [(3N-4-TIPOiPr)Co(H2O)(2O-NO3)]NO3 and [(3N-4-TIPOiPr)Cu(H2O)(O-SO4)], 4-TIPOiPr = tris(2-isopropylimidazol-4(5)-yl)phosphaneoxide Z. Anong. Allg. Chem. 633: 955–960. doi: 10.1002/zaac.200700031

68. Fan Y, Gao YQ (2007) A DFT Study on the Mechanism of Phosphodiester Cleavage Mediated by Monozinc Complexes. J. Am. Chem. Soc. 129: 905–913. doi: 10.1021/ja0660251

69. Yin X, Lin C, Zhou Z, Chen W, Zhu S, et al. (1999) Studies of artificial hydrolytic metalloenzymes: The catalyzed carboxyester hydrolysis by copper(II), zinc(II) and cobalt(II) complexes of the tripod ligand tris(2-benzimidazylmethyl)amine. Transit. Metal. Chem. 24: 537–540.

70. Nakata K, Shimomura J, Shiina N, Izumi M, Ichikawa K, et al. (2002) Kinetic study of catalytic CO2 hydration by water-soluble model compound of carbonic anhydrase and anion inhibition effect on CO2 hydration. J. Inorg. Biochem. 89: 255–266. doi: 10.1016/s0162-0134(01)00419-6

71. Mauksch M, Brauer M, Weston J, Anders E (2001) New Insights into the Mechanistic Details of the Carbonic Anhydrase Cycle as Derived from the Model System [(NH3)3Zn(OH)]+/CO2: How does the H2O/HCO3− Replacement Step Occur? ChemBioChem 2: 190–198. doi: 10.1002/1439-7633(20010302)2:3<190::aid-cbic190>3.3.co;2-z

72. Kimura E, Koike T, Toriumi K (1988) A Trigonal Bipyramidal Zinc(II) Complex of Phenol-pendant Macrocyclic Triamine. Inorg. Chem. 27: 3687–3688. doi: 10.1021/ic00293a056

73. Koziol L, Valdez CA, Baker SE, Lau EY, Floyd, III WC, et al. (2012) Towards a small molecule, biomimetic carbonic anhydrase model: theoretical and experimental investigations of a panel of Zinc(II) aza-macrocyclic catalysts. Inorg. Chem. 51: 6803–6812. doi: 10.1021/ic300526b

74. Brown RS, Curtis NJ, Huguet J (1981) Tris(4,5-diisopropylimidazol-2-yl)phosphine:zinc(2+). A catalytically active model for carbonic anhydrase. J. Am. Chem. Soc. 103: 6953–6959. doi: 10.1021/ja00413a031
75. Hakansson K, Wehnert A, Liljas A (1994) X-ray analysis of metal-substituted human carbonic anhydrase II derivatives. Acta Crystallogr. Sect. D 50: 93–100. doi: 10.1107/s0907444993008790
76. Simonsson I, Jonsson BH, Lindskog S (1979) A 13C nuclear magnetic resonance study of CO2/HCO3- exchange catalyzed by human carbonic anhydrase C at chemical equilibrium. Eur. J. Biochem. 93: 409–417. doi: 10.1111/j.1432-1033.1979.tb12837.x
77. Krebs JF, Ippolito JA, Christianson DW, Fierke CA (1993) Structural and functional importance of a conserved hydrogen bond network in human carbonic anhydrase II. J. Biol. Chem. 268: 27458–27466.
78. Merz KM, Banci L (1997) Binding of Bicarbonate to Human Carbonic Anhydrase II: A Continuum of Binding States. J. Am. Chem. Soc. 119: 863–871. doi: 10.1021/ja963296a
79. Behravan G, Jonsson BH, Lindskog S (1990) Fine tuning of the catalytic properties of carbonic anhydrase. Studies of a Thr200--His variant of human isoenzyme II. Eur. J. Biochem. 190: 351–357. doi: 10.1111/j.1432-1033.1990.tb15582.x
80. Fisher SZ, Maupin CM, Budayova-Spano M, Govindasamy L, Tu CK, et al. (2007) Atomic crystal and molecular dynamics simulation structures of human carbonic anhydrase II: insights into the proton transfer mechanism. Biochemistry 46: 2930–2937. doi: 10.1021/bi062066y
81. Loferer MJ, Tautermann CS, Loeffler HH, Liedl KR (2003) Influence of backbone conformations of human carbonic anhydrase II on carbon dioxide hydration: Hydration pathways and binding of bicarbonate. J. Am. Chem. Soc. 125: 8921–8927. doi: 10.1021/ja035072f
82. Bertini I, Canti G, Luchinat C, Mani F (1981) pH-Dependent properties of a CoN4(OH2) chromophore: a spectroscopic model of cobalt carbonic anhydrase. Inorg. Chem. 20: 1670–1673. doi: 10.1021/ic50220a010

83. Benelli C, Bertini I, Di Vaira M, Mani F (1984) Single-crystal x-ray and spectroscopic studies on the complex aquo(tris[(3,5-dimethyl-1-pyrazolyl)methyl]amine)cobalt(II)perchlorate. A spectroscopic model of cobalt-substituted carbonic anhydrase. Inorg. Chem. 23: 1422–1425. doi: 10.1021/ic00178a024

84. Davy R (2009) Development of catalysts for fast, energy efficient post combustion capture of CO_2 into water; an alternative to monoethanolamine (MEA) solvents. Energy Procedia 1: 885–892. doi: 10.1016/j.egypro.2009.01.118

85. MacAuley SR, Zimmerman SA, Apolinario EE, Evilia C, Hou YM, et al. (2009) The archetype gamma-class carbonic anhydrase (Cam) contains iron when synthesized in vivo. Biochemistry 48: 817–819. doi: 10.1021/bi802246s

86. Alber BE, Colangelo CM, Dong J, Stalhandske CMV, Baird TT, et al. (1999) Kinetic and spectroscopic characterization of the gamma-carbonic anhydrase from the methanoarchaeon Methanosarcina thermophila. Biochemistry 38: 13119–13128. doi: 10.1021/bi9828876

87. Iverson TM, Alber BE, Kisker C, Ferry JG, Rees DC (2000) A closer look at the active site of gamma-class carbonic anhydrases: high-resolution crystallographic studies of the carbonic anhydrase from Methanosarcina thermophila. Biochemistry 39: 9222–9231. doi: 10.1021/bi000204s

88. Innocenti A, Zimmerman S, Ferry JG, Scozzafava A, Supuran CT (2004) Carbonic anhydrase inhibitors. Inhibition of the beta-class enzyme from the methanoarchaeon Methanobacterium thermoautotrophicum (Cab) with anions. Bioorg. and Med. Chem. Lett. 14: 3327–3331. doi: 10.1016/j.bmcl.2004.06.073

89. Zastrow ML, Peacock AFA, Stuckey JA, Pecoraro VL (2011) Hydrolytic catalysis and structural stabilization in a designed metalloprotein. Nature Chemistry 4: 118–123. doi: 10.1038/nchem.1201

Chapter 7

ESR Study of Interfacial Hydration Layers of Polypeptides in Water-Filled Nanochannels and in Vitrified Bulk Solvents

Yei-Chen Lai[1], Yi-Fan Chen[2], and Yun-Wei Chiang[1]

[1]Department of Chemistry and Frontier Research Center on Fundamental and Applied Sciences of Matters, National Tsing Hua University, Hsinchu, Taiwan

[2]Department of Chemical and Materials Engineering, National Central University, Jhongli, Taiwan

ABSTRACT

There is considerable evidence for the essential role of surface water in protein function and structure. However, it is unclear to what extent the hydration water and protein are coupled and interact with each other. Here, we show by ESR experiments (cw, DEER, ESEEM, and ESE techniques) with spin-labeling and nanoconfinement techniques that the vitrified hydration layers can be evidently recognized in the ESR spectra, providing nanoscale understanding for the biological interfacial water. Two peptides of different secondary structures and lengths are studied in vitrified bulk solvents and in water-filled nanochannels of different pore diameter (6.1~7.6 nm). The existence of surface hydration and bulk shells are demonstrated. Water in the immediate vicinity of the nitroxide label (within the van der Waals contacts, ~0.35 nm) at the water-peptide interface is verified to be non-crystalline at 50 K, and the water accessibility changes little with the nanochannel dimension. Nevertheless, this water accessibility for the nanochannel cases is only half the value for the bulk solvent, even though the peptide structures remain largely the same as those immersed in the bulk solvents. On the other hand, the hydration density in the range of ~2 nm from the nitroxide spin increases substantially with decreasing pore size, as the density for the largest pore size (7.6 nm) is comparable to that for the bulk solvent. The results demonstrate that while the peptides are confined but structurally unaltered in the nanochannels, their surrounding water exhibits density heterogeneity along the peptide surface normal. The causes and implications, especially those involving the interactions between the first hydration water and peptides, of these observations are discussed. Spin-label ESR techniques are proven useful for studying the structure and influences of interfacial hydration

INTRODUCTION

Water is essential for the stability and function of biological macromolecules, such as proteins, and DNA. Recent studies have investigated water in the first few hydration layers of proteins, the so-called "biological surface water". [1]–[3] Evidence has been gathered to suggest that the properties of this surface water shell are extremely important to biological functioning of a protein; [4], [5] on the contrary, some studies have reported weak interdependence between the interfacial hydration and the protein structure, where the latter is of tremendous influence to its functioning. [6], [7] the role of the interfacial hydration therefore remains a matter of debate. To resolve this ambiguity, a molecular scale picture of how water and protein structurally and functionally interact with each other, and why the first surface hydration water differs from bulk water is much needed but still largely lacking. This absence severely impedes our understanding toward the biological water. A reliable tool that has sufficient sensitivity and accuracy to probe the changes, at a molecular level, in dynamics and structure for both protein and water molecules has been long desired. [1], [8] Recently, spin-label electron spin resonance (ESR) has been demonstrated useful for probing water dynamics. [9], [10] Built upon this progress, an approach combining ESR with nanochannels has emerged as a potentially useful tool for studying protein-water interactions; it demonstrated that the nanochannel of mesoporous materials was useful to maintain an amorphous state of solvents, which is required for studying protein structures and their dependence on water properties. [11], [12] Moreover, studying the biological surface water within nanochannels *per se* is also of fundamental importance in nature, since it is related to the origins of life. [13] Here, we report a carefully comparative investigation on the surface water molecules that directly interact with the biomolecules in nanochannels versus those for bulk solvents, and demonstrate that the surface hydration layer can be evidently recognized in the ESR spectroscopy.

In this work, we studied the structural conformations and local solvation of a 26-residue polypeptide (n3) and a 14-proline-long model peptide (PPm3), as immersed in bulk solvents versus encapsulated within the nanochannels of mesoporous materials, by several of the pulsed ESR techniques including double electron-electron resonance (DEER), electron spin echo (ESE), and electron spin echo envelope modulation (ESEEM) experiments. The n3 and PPm3 peptides in pure water have a propensity to a -hairpin structure and a left-handed polyproline II (PPII) helix conformation, respectively. In the spin-label ESR studies, [14]–[16] the introduction of a cysteine-specific paramagnetic nitroxide probe is accomplished through cysteine substitution mutagenesis; for peptide, cysteine is introduced during synthesis. The DEER technique [17], [18] allows a careful examination of the differences in the biomolecular structures before/after the encapsulation into the nanochannels. [11] The ESE and ESEEM are two powerful tools capable of reporting the coupling between a nitroxide spin and nearby water protons within the ranges of ~2 nm and ~0.6 nm, respectively. [19], [20]. We show by DEER technique that the two peptides retain their secondary structures after the encapsulation into the nanochannels as well as in a mostly vitrified bulk solvent at cryogenic temperatures. In the further ESE and ESEEM measurements, we carefully examine the density of the interfacial water within the range of 2 nm from the peptide surface. The inner (i.e., the first hydration shell) and outer (the bulk-like shell) water shells respond differently regarding the nanochannel diameters and with/without nanoconfinement. The differences are quantitatively described by the water density derived from the ESE and ESEEM measurements. The details regarding the local structural changes of the water/peptide molecules are thus unambiguously revealed based on the results of the ESR spectroscopy.

MATERIALS AND METHODS

Polypeptides

All peptides of this study were custom-synthesized by Kelowna International Scientific Inc. (New Taipei, Taiwan) with purity greater than 95%. The n3 peptide is a linear 26-mer polypeptide with a sequence of GN<u>DYEDRYYRENM</u>YRYPN<u>QVYYR</u>PVA, where the first 25 residues correspond to the domain 142–166 of the human prion protein and the underlined letters represent the segments that correspond to helix H1 and β-strand S2, respectively, in the normal prion structure. [21] It is highly soluble and exhibits an intrinsic propensity to a β-hairpin conformation at neutral pH in the PB buffer (10 mM sodium phosphate, pH 6.5) or in pure water. [12] It was previously demonstrated that the n3 peptide folds into an α-helical conformation in a TFE (trifluoroethanol)/PB buffer. [12], [22] Two mutations of the n3 peptide were studied. They were substituted with a cysteine at the 9th site and at both the 3rd and 9th sites, respectively, and were denoted as n3-s and n3-d (cf. Figures 1a and 2a for the structures of the n3β-d and n3α-d, respectively). Pure H_2O and D_2O were used in the n3-s study by ESEEM. The PPm3-s is a 14-mer-long polyproline model peptide, with the 9th residue substituted with cysteine. Previously this PPm3-s was shown to remain a PPII helical structure as immersed in a vitrified bulk solvent (40% sucrose in water) or encapsulated into the mesoporous materials containing pure water.

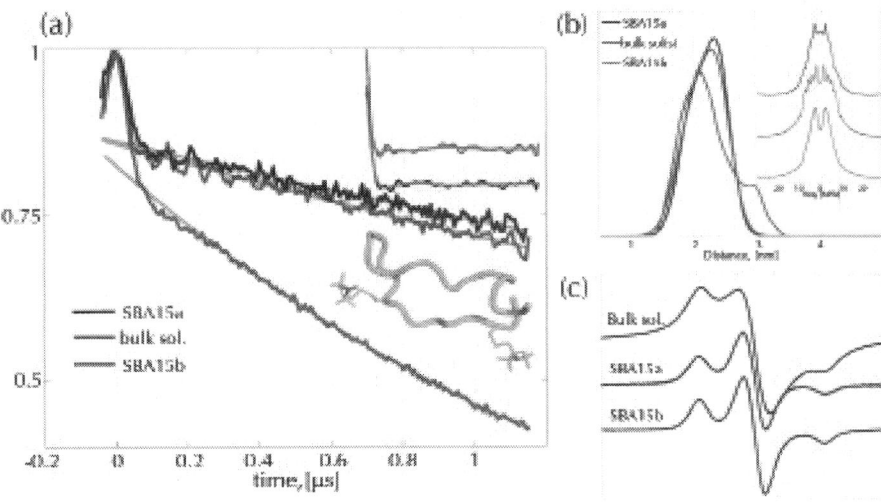

Figure 1: Determination of the n3β-d structure. (a) The time-domain DEER data for the n3β-d (0.5 mM) in the studied conditions, including the vitrified bulk solvent (sol(s)/H_2O) and the nanochannels (SBA15a and SBA15b). The gray lines represent the exponential baselines that best fit the DEER data. There are two insets. One displays a ribbon model for the n3β-d showing the spin-label side-chains at the 3rd and 9th sites of the peptide. The model was derived from a NMR study (PDB code: 1G04). The other inset shows the baseline-corrected DEER traces for the sol(s)/H_2O and the SBA15a, and also the simulated DEER traces (in green color) using the obtained P(r)s. There are some distinct differences in the two traces. (b) The (normalized) interspin distance distributions of the n3β-d peptides in the conditions studied. The average distances of the three measurements are approximately the same, indicating the n3β structure remains roughly unchanged. A much-broadened P(r) for the bulk solution study is obtained due to the solvent heterogeneity. The inset shows the Pake doublets converted from the DEER data. (c) Cw-ESR spectra of the n3β-d at 50 K. The clustering, caused by the solvent heterogeneity at 50 K, is evidently observed in the cw-ESR spectra of the bulk solution study, but not in the nanochannel studies.

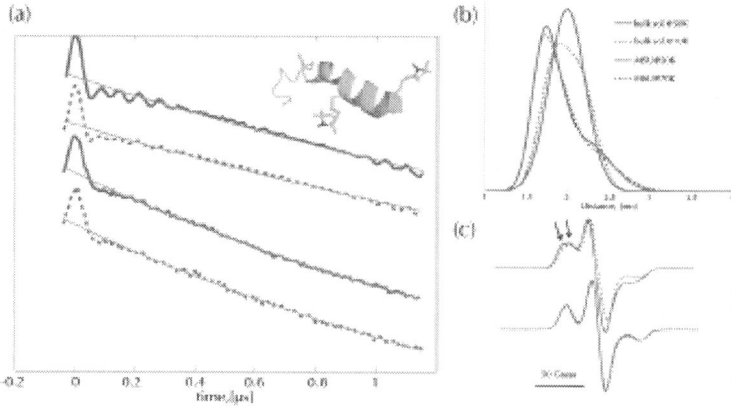

Figure 2: Determination of the n3α-d structure. (a) The time-domain DEER signals of the studied conditions. The gray lines represent the exponential baselines that best fit the data. Inset shows a ribbon model of the n3α-d derived from NMR data (PDB code: 1M25). (b) The P(r) distributions extracted from the time-domain DEER data by the Tikhonov regularization analysis. The average distances (~2.0 nm) are consistent with the expectation. (c) The cw-ESR spectra of the n3α-d. The spectra of the bulk solution studies are characterized by a broader linewidth and the spectral heterogeneity (indicated by arrows) as compared to the spectra of the nanochannel studies.

The circular dichroism (CD) spectroscopy measurements for the n3-s and n3-d peptides as in the nanochannels versus in the bulk solvent (at 4°C and 25°C) were previously reported in Huang et al [12] and shown to resemble each other closely showing a typical β-hairpin structure (or an α-helical structure in the TFE/PB solvent). The results indicated that the secondary structures of the studied n3 peptide variants remain unchanged over the investigated experimental conditions [23].

Nanochannels

The mesoporous silica SBA15 materials were synthesized as previously described. [24] Briefly, tetraethoxysilane was added to the HCl solution of triblock copolymer Pluronic P-123 ($EO_{20}PO_{70}EO_{20}$). The molar composition was 1 TEOS : 0.54 HCl : 100 H_2O : 0.017 P-123. The mixture was stirred at 35°C for 24 h, aged at 90°C (SBA15a) and 60°C (SBA15b), for 24 h and then filtered and dried. The copolymer molecules in the as-synthesized SBA15 samples were removed by treating the samples with concentrated sulfuric acid at 90°C followed by calcination at 350°C. The structural properties of the mesoporous materials are summarized in the following order: SBA15a, SBA15b. The pore diameters are 7.6 and 6.1 nm. The unit cell sizes are 11.6 and 9.6 nm. The sizes of the wall thickness are thus 4.0 and 3.5 nm. The nanochannel used for the n3α-d study was a mesoporous silica material with a cross-linked hexagonal pore structure and an average pore size of 7.1 nm (known as MSU), and was purchased from Aldrich. The MSU nanochannel was demonstrated useful for encapsulating spin-labeled peptides for ESR study. [11] The smallest pore size (6.1 nm) for the DEER measurements of the present study is at least 1.5 times larger than the globular sizes of the n3 and PPm3 peptides. In DEER measurements, different nanochannels would result in a change in the slope of the signal baseline, but give no rise to changes in the secondary structures.

Experimental Procedures

In the spin-labeling experiment, peptides were labeled with a 10-fold excess of (1-Oxy-2,2,5,5-tetramethyl-3-pyrroline-3 -methyl) methanethiosulfonate spin label (MTSL) (Alexis biochemicals, San Diego, CA) per cysteine residue for overnight in the dark at 4°C. They were further purified by reverse phase HPLC as previously described. [12] MALDI-TOF experiments were conducted to confirm the identity of the peptides carrying the spin labels. In the bulk solution studies, two cryoprotectants (sucrose and glycerol) were used. The concentration of the spin-labeled peptides was 0.5 mM for n3β-d; the concentration was 1.5 mM for the studies of n3β-s and PPm3-s. We used the n3-d in the DEER measurements and n3-s as well as PPm3-s in the ESEEM and ESE measurements. For the n3α studies, the concentration of the n3 -d was 0.9 mM in the (mostly) vitrified solution of the 40% (w/w) sucrose and 87/13 (v/v) TFE/PB mixture. The n3α-d concentration for the nanochannel study was 1.2 mM in the solvent of 87/13 (v/v) TFE/PB. The solution volume added into an ESR tube for the bulk solution studies was ca. 40 μL. A higher percentage of the cryoprotectant in solution was rarely reported, and was not recommended for biological experiments as it could easily cause disturbance to the native conformation of biomolecules. The encapsulation of the samples into the nanochannels was prepared as previously described. [11], [12], [25] The preparation approach has proven useful for effectively trapping the added solution within the nanochannels. The solution volume added onto the mesoporous materials (0.1 mL; 12 mg) was ca. 20 µL. After the encapsulation, the outside surface of the nanochannel materials appeared dried. No cryoprotectant was used for the nanochannel experiments. To perform the deuterium-hydrogen exchange experiment to modify the surface group of the nanochannel materials, a previously developed procedure was followed. [26] The materials were placed in a vacuum oven at 100°C overnight to remove moisture. Dried mesoporous materials were submerged into pure excess D_2O at room temperature. After >1.5 hrs, the excess D_2O was pipetted out and the sample was refreshed with fresh D_2O. The cycle was repeated four times, followed by an overnight incubation. The sample was dried again before the peptide encapsulation.

In summary of the studied solvent conditions, sol(s)/H_2O represents a mixture of a bulk solvent containing 40 wt% sucrose and H_2O; as D_2O is used, it is denoted by sol(s)/D_2O. Sol(g)/H_2O stands for a mixture of a bulk solvent containing 40v/v% glycerol and H_2O, while sol(dg)/D_2O denotes a mixture of a bulk solvent containing 40v/v% deuterated glycerol and D_2O.

Cw/Pulsed ESR Measurements and Data Analysis

A Bruker ELEXSYS E580 cw/pulsed spectrometer, equipped with a Bruker pulse ELDOR unit E580-400, a dielectric resonator (ER4118X-MD5W), and a helium gas flow system (4118CF and 4112HV), was used. DEER experiments were carried out using the four-pulse constant-time DEER sequence: $\pi/2(\omega_A)-\tau_1-\pi(\omega_A)-(\tau_1+t)- \pi(\omega_B)-(\tau_2-t)-\pi(\omega_A)-\tau_2$-echo. The detection pulses (ω_A) were set to 32 and 16 ns for π and $\pi/2$ pulses, respectively and pump frequency (ω_B) was set to approximately 70 MHz lower than the detection pulse frequency. The pulse amplitudes were chosen to optimize the refocused echo. The $\pi/2$-pulse was employed with +x/−x phase cycle to eliminate receiver offsets. The duration of the pumping pulse was about 32 ns, and its frequency (ω_B) was coupled into the microwave bridge by a commercially available setup from Bruker. All pulses were amplified via a pulsed traveling wave tube amplifier (E580–1030). The field was adjusted such that the pump pulse is applied to the maximum of the nitroxide spectrum, where it selects the central $m_I = 0$ transition of A_{zz} together with the $m_I = \pm 1$ transitions. The echo was measured as a function of t, while τ_2 was kept constant, depending on T_M. Typical numbers of shots per points and scan number were set to 1024 and 30–40, respectively. Accumulation time for each set of data was about 12 hours. Most of the pulsed ESR experiments of this study were performed at 50 K while some were performed at 70 K. The employed cooling scheme is described below. The determination of interspin distance distribution of the DEER data was performed in the time-domain analysis by Tikhonov regularization based on the L-curve method. [27], [28] This four-pulse DEER experiment has become a widely used approach for measuring distances between electron spins (or spin labels) in proteins and protein/membrane systems in the range approximately 1.5 to 8 nm. [29] The cw-ESR experiment was performed at an operating frequency of 9.4 GHz and 1.5 mW incident microwave power. The swept magnetic range was 200 Gauss.

The three-pulse ESEEM is a powerful technique to measure weak coupling between paramagnetic ions and nearby (<0.6 nm) nuclei with nonzero spin metals in biological systems, such as metalloproteins and bioinorganic compounds. [20], [30] Its sequence was $\varpi/2-\tau-\varpi/2-T-\varpi/2-\tau$–Stimulated Echo with a pulse length $t_{\varpi/2} = 16$ ns, the T value starting from 400 ns with a time increment $\Delta T = 4$ ns. The data were collected by monitoring the amplitude of the stimulated echo as the time T was increased by ΔT. In order to avoid the blind spot and to ensure that all frequencies are being detected, signals were collected by varying the time τ from 100 to 400 ns. The τ values of 260 and 100 ns were selected in the τ variation experiments and used for three-pulse ESEEM measurements. A four-step phase cycling was used to remove unwanted echoes. The time-domain ESEEM data were first baseline corrected by a stretched exponential, zero filled and afterwards fast Fourier transformed. The FT-ESEEM spectra were shown in absolute magnitude. All the theoretical simulations and analyses were performed using MATLAB with EasySpin toolbox [31].

ESE is a useful tool to provide information concerning the interaction between an unpaired spin and neighboring electron and nuclear spins approximately within a 2-nm range. The ESE experiments were applied using the 2-pulse Hahn echo sequence, consisting of a $\varpi/2$ pulse along the x-axis followed by a delay and a train of ϖ pulses, separated by interpulse delays 2 . [32] The field was adjusted to optimize the spin echo, and the duration times of $\varpi/2$ and ϖ pulses were set to 16 and 32 ns. As previously described, [11] the ESE signals were fitted to a stretched exponential function to extract T_M value from the ESE data.

A common cooling approach was used. The sample tube was plunge-cooled in liquid nitrogen, and then transferred into the ESR probehead, which was pre-cooled to 50 K using the helium flow system. This rapid cooling scheme is supposed to preserve the room-temperature molecular features of all relevant molecules at cryogenic temperatures, as premised in the X-ray protein cryo-crystallography [33]; the molecular conformations observed at cryogenic temperatures are thus *snapshots* of what had happened at room-temperature. [34], [35] Further discussion regarding this aspect is presented below. To test the consistency of the cooling procedure, some measurements (e.g. the n3 studies) were repeated three times, by cooling different samples taken from the same stock solution. Same cooling procedure was followed in all of the measurements to ensure that the cooling rate is approximately the same. The results were highly reproducible.

Experimental Verifications of the Peptide Encapsulation and Preservation of the Peptide Properties upon Encapsulation

To verify the peptides are adequately encapsulated into the nanochannels, this study performed the following experiments. An excess buffer was added into the ESR tube containing the mesoporous materials, wherein the spin-labeled peptides were encapsulated as described above. The tube was sent for ESR measurements at room temperature. The collected spectra were found to remain identical to the spectra collected before the addition of the excess buffer; that is, the spectra show typical slow-motional lineshapes. This experiment demonstrated that the spin-labeled peptides were not adsorbed/left on the outer surface of the materials and that the molecules were trapped adequately well within the nanochannels. This study also confirmed that no ESR signal was obtained for the supernatant liquid after centrifugation.

In all of the nanochannel studies, no cryoprotectant was used. At such a low temperature (50 K), pure water became ice. In one control experiment, this study obtained noise-like signals in the pulsed ESE measurements for the n3 in the bulk water without the cryoprotectant. This control experiment supported the following. In the nanochannel experiments, the tiny amount of the solution that might be left outside the mesopores would become ice crystals, resulting in an extremely short spin phase memory time (T_M). Accordingly, the rapidly decaying signals (due to the short T_M value) would by no means be observed in our pulsed ESR experiments. Therefore, the meaningful pulsed ESR signals that we collected at 50 K were absolutely from the samples within the nanochannels.

Previously, [11], [12], [25] we studied the n3 and n3 peptides by cw-ESR. We demonstrated that the spectra of the n3 and n3 peptides were distinctly different as confined in the nanochannels of the same material, and that the spectra varied reasonably concerning temperature variations (200 ~ 300 K). Besides, it was confirmed by the CD and Pulsed ESR experiments that the secondary structures of the n3 peptides were not disrupted in the nanochannels. [12] Same observations were made for the PPm3, showing that the PPm3 retained its PPII structure in the nanochannels. [23] These observations evidently indicated that the interaction between spin label and the inner surface of the nanochannel, if any, would definitely be insignificant as compared to the backbone dynamics of the secondary structures and the interactions between solvent molecules and the peptides. Otherwise, the spectra of the n3 peptides in helical versus hairpin forms would have appeared similar. In view of the observations of the experiments described above, it is shown that the peptides are immersed in solvent and are confined but free as encapsulated within the nanochannels.

RESULTS

Determination of n3β Peptide Structure by DEER

Two different nanochannels, SBA15a and SBA15b with pore diameters of 7.6 and 6.1 nm, respectively, are used. In the bulk solution experiments, the (mostly) vitrified bulk solvent is 40 wt% sucrose in H_2O (or D_2O), denoted by sol(s)/H_2O or sol(s)/D_2O. In the nanochannel experiments, the solvent contains no cryoprotectant (sucrose).

The time-domain DEER traces for the three studied conditions (with the same nominal spin concentration of 0.5 mM) are shown in Figure 1(a), wherein a cartoon model of the n3β-d peptide is shown as an inset. The baseline for the SBA15b decays rapidly as compared to the baselines of the SBA15a and bulk solution studies, indicating that the bulk spin concentration is increased upon the encapsulation into the SBA15b. The effect of the increased concentration in the SBA15b will be discussed later. A typical DEER analysis procedure was performed to remove the baselines with a homogeneous model. [36] As the time-domain traces for the sol(s)/H_2O (blue) and SBA15a (black) look somewhat similar in Figure 1(a), we plot their respective normalized baseline-corrected DEER traces in the inset (upper right) of Figure 1(a), wherein the green lines represent the simulated DEER traces using the P(r) shown in Figure 1(b). The two traces in the inset are shifted by 0.05 in y-axis for clarity. The inset plot clearly shows that there are small but significant differences in the two traces. Figure 1(b) reports the intra-molecular interspin distance distributions P(r), determined by the Tikhonov analysis, [27] for the DEER time-domain data after the baseline removal. The P(r) results are normalized to have the same area. The two distance distributions of the nanochannel studies are alike, showing the same average distance of ~2.2 nm. It is consistent with the structure and backbone flexibility of the n3 peptide derived from the NMR data taken at 5–37 °C. [21] Conducted at 200~300K, a previous study demonstrated that the side-chain disordering (i.e., the cone width of the side-chain) of the spin-labeled n3, as confined in nanochannels, decreased substantially as compared to the disordering in the bulk solvent. [12] Therefore, the obtained P(r) results of the confined n3-d largely reflect the backbone conformation and flexibility that are closely correlated with the local solvent heterogeneity (although the contribution from the side-chain rotamers cannot be completely excluded). For the bulk solution study, the result shows a most probable distance around 2.1 nm and a minor peak at ~3 nm. This heterogeneous distribution is in accordance with a previous finding [11] showing a broadened trimodal-like P(r), with peaks at 2.0, 2.2, and 2.9 nm, for the same solvent and the same n3-d peptide at 50 K. Taken together, this heterogeneity in the P(r) for the n3-d can be consistently observed for the bulk solvent case in different measurements. To confirm the P(r) results, we perform Fourier transformation to convert the DEER data to Pake doublets as shown in inset of Figure 1(b). The splitting of the Pake doublets is distinctly less obvious in the bulk solution data than in the nanochannel results, validating the existence of the 3-nm distance peak in the P(r) of the bulk solution study. Moreover, this study verified the existence of the 3-nm peak with a newly developed method [37] and confirmed that the peak is not a ghost peak and cannot be removed by adding noise to the time-domain data. The cw-ESR spectra collected at 50 K (Figure 1c) shows that a large spectral broadening, which is caused by the peptide clustering due to the partial ice crystal formation in the solvent [11], can only be observed in the spectra of the bulk solution study but not in the nanochannel studies. This clustering of the n3 in the bulk solvent was previously reported [11] and demonstrated in a spin dilution experiment, which also showed that the spectral broadening could be significantly reduced with a decrease in the concentration of the spin-labeled n3 in the bulk solvent. Collectively, as indicated by the similarity of the P(r) distributions and the broadened cw-ESR spectral features, this study concludes that i) the n3 structure remains approximately (~2.2 nm) unchanged over the studied conditions, and ii) the peculiar 3-nm peak is a composite result of the partial water crystallization within the solvent upon cooling to 50 K, the so-called cryoartifact.

Determination of n3α Peptide Structure by DEER

The need for further confirming the structural integrity of n3 in our studied conditions is warranted since the DEER is performed at 50 K and the determined distances (Fig. 1) are along one leg of the n3-d, even though the CD results in the previous study conducted at room temperature exhibit the same spectra as reported for the n3. We therefore study the structure of n3-d both in the bulk solvent and within the nanochannels at 50K.

Inset in Figure 2(a) shows a cartoon model of the n3-d peptide. The time-domain DEER traces for the n3-d in the (mostly) vitrified bulk solvent (blue) and in the MSU nanochannel (red) at temperatures 50 K (solid lines) and 70 K (dashed lines) were shown in Figure 2(a). The gray lines represent the exponential baselines that best fit the time-domain DEER data. The time-domain DEER signals displayed in Figure 2(a) were equally y-shifted for a clear presentation of the experimental data. Figure 2(b) reports the intra-molecular interspin distance distributions P(r), determined by the Tikhonov analysis [28], for the time-domain DEER data after the baseline removal. For the MSU nanochannel studies, the P(r) results resemble each other closely, indicating that the structure remains approximately the same over the temperature range. The most probable distance and the full-width at half-maximum (fwhm) of the peak are ca. 2.0 nm and 0.51 nm, respectively. The C-C distance in the determined NMR structure[22] (PDB code: 1M25) of the n3 peptide is 1.1 nm, showing that the side chains of the two studied sites stretch out into the solvent in opposite directions from the helix axis (cf. Fig. 2a). Since the stretch length of one spin-label side-chain could be up to 0.7 nm, the determined distances (~ 2.0 nm) of the present study are reasonably consistent with the structure of the NMR study. In the bulk solution studies, the P(r) results of 50 and 70 K are closely similar. The average distance is about 1.93 nm. As compared the P(r) results of the bulk solution to those of the MSU, they are slightly different wherein the former is clearly broader than the latter. The P(r) results of the bulk solution appear to have a small distance shoulder around 2.4 nm in addition to the major distance at 1.75 nm. The P(r) heterogeneity for the bulk solution study can be observed in different measurements at varying temperature. We attribute the presence of the broad and somewhat of heterogeneity P(r)s to the consequence of the solvent heterogeneity (caused by partial water crystallization; cryoartifact), as the fwhm of the P(r) can be easily compromised by the side chain disordering and the fact that the averaged distances of the P(r)s are about the same. This observation for the P(r) is also supported by the cw-ESR results (cf. Figure 2c) wherein the spectra display the existence of the multiple spectral components. The low-field peak in the cw-ESR spectra of the bulk solution studies clearly shows the evidence of multiple spectral components (as indicated by arrows), manifesting the existence of the local heterogeneity in the solvent when cooled to low cryogenic temperatures, thereby resulting in the peculiar peak around 2.4 nm in the P(r)s which cannot be regarded as a ghost peak and removed in the DEER analysis. [37] This indicates that the interpeptide distances are within the cw-ESR sensitive region, i.e. <2 nm, displaying a sign of molecular clustering in the bulk solution upon cooling. Collectively, the average distances of the n3 -d are close in all of the studied conditions and the structure of the n3 -d remains intact upon the encapsulation into the nanochannels, although the solvent heterogeneity does give rise to ambiguity in the determined P(r). This conclusion is consistent with the results of the n3 -d (cf.Figure 1) and supports our claims made in the end of last section.

Interfacial Water Accessibility by ESEEM

To unravel the details of the local solvation of the encapsulated peptides in the nanochannels, this study carried out three-pulse ESEEM experiments at 50 K for [n3 -s] = 1.5 mM and [PPm3-s] = 1.5 mM in various local environments. ESEEM is a powerful technique to measure weak coupling between paramagnetic ions and nearby (<0.6 nm) nuclei with nonzero spin. [20], [30]Only magnetic nuclei that are weakly coupled to the spin label give rise to peaks in Fourier Transformed (FT)-ESEEM spectrum, centered approximately at the Larmor frequencies of the coupled nuclei. [38] Under the condition of strong coupling (i.e., $A_{iso} \geq 2\nu_I$, where A_{iso} is the isotropic value of hyperfine tensors, and ν_I the nuclear Larmor frequency), it would lead to a splitting away from the Larmor frequency [39].

Figure 3 shows (a) time-domain ESEEM and (b) the FT-ESEEM plots of the time-domain ESEEM for n3 -s. Note that the dashed lines in Figure 3(a) are theoretical fits to the experimental data and are to be discussed in the later paragraph. The studied conditions include a (mostly) vitrified bulk solvent and a variety of combinations varying nanochannel material (SBA15a or SBA15b), solvent (H_2O or D_2O), and surface groups inside the materials (-SiOH or -SiOD). The results clearly show that the peak position is dependent on the solvent molecules. As H_2O is used as the solvent, the peak in the FT-ESEEM spectra is observed at frequency 14.7 MHz, which is approximately the same as the Larmor frequency of nucleus ^1H. The peak in the FT-ESEEM spectra shifts to 2.2 MHz, which matches the Larmor frequency of deuterium, as D_2O is used as the solvent. This demonstrates that the coupling between the nitroxide spin and the solvent molecules is a weak coupling.

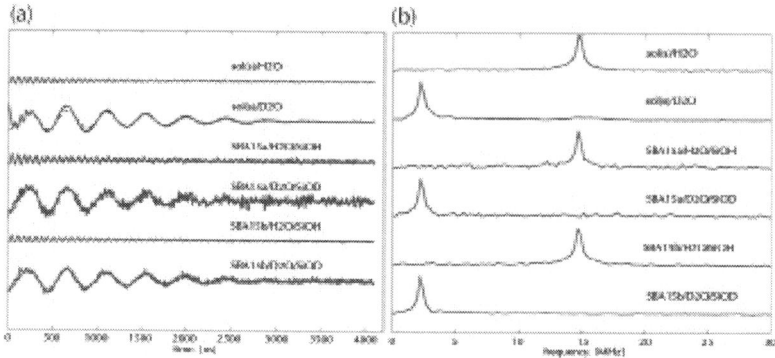

Figure 3: Water accessibility study of the n3 -s by ESEEM. (a) Three-pulse ESEEM time-domain data (solid lines) after the removal of the exponential decaying function in the raw data. The modulation depth is directly correlated to the peak intensity of the FT-ESEEM and can be quantitatively described by k_D and obtained in the following analysis. The dashed lines represent the theoretical fits to the experimental data. According to ESEEM theory, a damped harmonic oscillation

$$f(t) = k_D \cos(2\pi v_D t + \vartheta)\exp(-t^2/\tau_0^2),$$ as a function of the ESEEM collection time t, can be employed. [20] The $f(t)$ function was least-square fitted to the ESEEM time-domain data by varying the deuterium modulation depth, k_D, damping constant, τ_0, and phase, , while keeping the Larmor frequency of deuterium v_D fixed. The best-fit parameters are shown in Table 1. The signal-to-noise ratio of the experiments varies with many experimental conditions, such as instrumental setups, data collection time, and so on. Thus, FT-ESEEM has been a typical data presentation for the ESEEM analyses. (b) The FT-ESEEM data for the n3 -s in the studied conditions, including the vitrified bulk solvent and a variety of combinations of the nanochannel materials (SBA15a, SBA15b), solvents (H_2O, D_2O), and surface groups inside the materials (-SiOH or -SiOD).

In the bulk solution study of sol(s)/D_2O (Fig. 3), the FT-ESEEM spectrum displays only the peak for deuterium but not hydrogen. This suggests that the solvent molecules within the ESEEM sensitive region (ca. <0.6 nm) near to the nitroxide spin are mostly D_2O or a mixture of D_2O molecules and OD of sucrose after the D/H exchange with the D_2O bath. Note that the intensity of ^2H in the FT-ESEEM representation is at least 20 times greater than that of ^1H.

According to ESEEM theory, the modulation depth can be easily extracted from time-domain ESEEM signals, and is dependent on the number and distances of the nuclei coupled to the electron spin. The data can be immediately analyzed to yield a three-pulse ESEEM-based (deuterated) water accessibility parameter, Π (D_2O), using Eq. (1), [19], [20]

$$\Pi(D_2O) = \frac{2k_D}{[1-\cos(2\pi v_D \tau)]} \times \left[\frac{v_D}{2MHz}\right]^2 \quad (1)$$

Where k_D and $_D$ are, respectively, the modulation depth and Larmor frequency of deuterium, and the choice of the first interpulse delay. Note that the k_D value can be precisely determined in a typical time-domain ESEEM analysis. [20] This parameter is sensitive to the presence of deuterium nuclei within the van der Waals contact distance of ~0.35 nm from the spin label. The obtained values for the n3-s are close (with small differences) to each other (cf. Table 1). Our finding shows that the water accessibility ( ) values of the spin-labeled site on the n3-s are only slightly greater for the SBA15a than for the SBA15b and the bulk solvent.

Table 1: Parameters obtained in the analyses of the ESE and ESEEM data[§]

D_2O	T_M (ns)[#]		$X^{\ddagger}$		$C_{ex}(nm^{-3})$	k_D	Π
	H_2O	D_2O H_2O D_2O			D_2O	D_2O	
n3 β-s-a	2950	1967	0.93	0.81	24.2	0.129	0.185
n3 β -s-b	2156	1455	0.93	0.79	42.3	0.109	0.156
n3 β -s-sol(s)	3545	2693	1.03	1.15	12.8	0.099	0.142

PPm3-s-a	2781	1785	0.82	0.81	28.6	0.151	0215
PPm3-s-b	1004	703	0.93	0.74	60.9	0.092	0.132
PPm3-s-sol(g)	7687	3430	0.95	1.52	23.0	0.335	0.479
PPm3-s-sol(s)	5192	3496	1.20	1.45	13.4	0.160	0.229

§Estimated errors: 5 %(TM), 10 % (x), 13% (C_{ex}), 5% (k_D), 10% (P). Abbreviations: n3-s-a (the n3-s is within SBA15a containing pure water); n_3-s-b (the n_3-s is within SBA15b containing pure water); n_3-s-sol(s) (the n_3-s is in a vitrified bulk solvent containing 40 wt.% sucrose, (s), in D_2O or H_2O); PPm3-s-sol(g) (PPm3-s is in a vitrified bulk solvent containing 40v/v% glycerol in H2O; deuterated glycerol is used if the solvent is D_2O, a condition of which is represented by sol(dg)/D_2O in main text); PPm3-s-sol(s) (PPm3-s is in a vitrified bulk solvent containing 40 wt% sucrose in D_2O or H_2O). In all of the experiments, the surface group of the nanochannels is modified to –SiOD in advance if D_2O is used. See Method for details.

#the values of T_M and x are obtained in the analysis of the pulsed ESE measurements using a stretched exponential function,

$$Y(2\tau) = Y(0)\exp[-(2\tau/T_M)^x],$$ where τ is the time between the two pulses, x the exponent, and Y(0) is the echo intensity at τ = 0. The obtained values are used to yield C_{ex} using Eq. (2). The C_{ex} represents ESE-based water accessibility within the range of, ~ 2 nm from the nitroxide spin.

"The k_D values are obtained in the theoretical analysis of the ESEEM measurements as described in Figure 3. The best-fit values for the damping constant (τ_0) and phase (ϕ) are very close together (2.9 ~ 3.0). The Π represents ESEEM-based water accessibility within the range of, ~ 0.35 nm from the nitroxide spin.

doi:10.1371/journal.pone.0068264.t001

Additionally, this study carries out the ESEEM measurements for the PPm3-s in various conditions as shown in Figures 4(a) and (b) for the time-domain versus FT-ESEEM, respectively. Two of the PPm3 variants (PP-3R1 and PP-2R1; see [23]), which are of the same length as PPm3 but differ in spin-labeling sites, were previously studied and demonstrated to possess a PPII helical structure in SBA15a, MSU, and in sol(s)/H$_2$O. The inset of Figure 4(b) shows the structure of the triply labeled PPm3 variant. In the present study, pure D$_2$O was used as the solvent and the inner surface of all the nanochannels was modified to be deuterated silanol groups in the measurements. The FT-ESEEM results clearly show the presence of the Larmor frequency of ^{2}H. Although the PPm3-s (in the sol(s)/D$_2$O) is supposed to be surrounded by a mixture of –OD and –OH, the signal corresponding to ^{1}H is not observed in the measurements. This is consistent with the observation for the n3 -s (cf. sol(s)/D$_2$O in Figure 3). By theoretical analysis (Eq. 1), the k_D and values can be extracted from the ESEEM measurements and used to determine the contributions solely from the deuteron. As shown in Table 1, except for the result of the PPm3-s with sol(dg)/D$_2$O, the values for the PPm3 studies are close to those obtained for the n3 studies and are within the typical range for a solvent-exposed site on a protein (discussed later). The differences in the values between the studies of the two peptides are small and are considered to reflect only the minor differences in the respective local solvation environments. Whereas, the k_D and values of the PPm3-s in sol(dg)/D$_2$O are at least two times greater than the others for the PPm3 (i.e. those without the deuterated cryoprotectants or without cryoprotectant at all in the nanochannel cases; cf. Fig. 4). Apparently, the deuterated glycerol molecules play an important role in the ESEEM signals and consequently the k_D and values. Therefore, this study shows that the water accessibility of the hydration layer at the interface with the peptides (within the ESEEM sensitive region) in the bulk solvents is at least two times greater than that in the nanochannels.

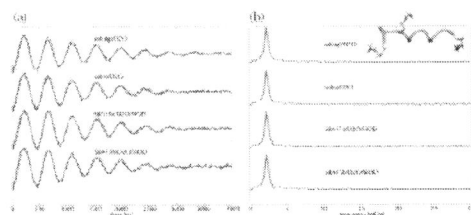

Figure 4: Water accessibility study of the PPm3-s by ESEEM. (a) Three-pulse ESEEM time-domain data (solid lines) after the removal of the exponential de-

caying function in the raw data. The modulation depth is directly correlated to the peak intensity of the FT-ESEEM and can be quantitatively characterized by the best-fit parameter k_D (cf. Table 1). The dashed lines represent the theoretical fits to the experimental data using the equation described in Figure 3. (b) The FT-ESEEM data for the PPm3-s in various deuterated conditions. The peaks correspond to the Larmor frequency of nucleus ^2H, indicating the PPm3-s is surrounded by D_2O. The inset shows a ribbon model of a PPm3 variant carrying three spin labels.

This analysis method was usefully demonstrated for obtaining a reliable water accessibility parameter for each of the spin-labeled sites in a large transmembrane protein in deuterated water. [20] Previously, value was found to be 0.13~0.25 for typical solvent-exposed sites on a protein, below 0.1 for buried residues, and >0.35 for unbound spin labels. [20] The obtained values of the present study (except for the sol (dg)/D_2O) are within the range for the solvent-exposed sites. They are distinctly greater than the values for a typical buried site of a protein, and less than the values for an unbound spin label. Our study indicates the following: i) the local solvation changes little with the secondary structures (n3 -s versus PPm3-s) as well as nanochannel sizes, and ii) the deuteron accessibility within the range ~0.35 nm from the spin label in the nanochannels is roughly half the value for the peptides immersed in the vitrified bulk solvents.

Interfacial Water is Non-crystalline

The interfacial water is expected to be amorphous (either liquid or solid) due to the rapid cooling scheme employed here, and the added cryoprotectants in the bulk case or the nanoconfinement in the nanochannel study. To confirm this expectation, this study also performed ESEEM measurement for the n3 -s (and DEER for the n3 -d) in pure water at 50 K, and found that no ESEEM (and DEER) signals were obtained, because of the formation of crystalline ice at 50 K, which resulted in a significantly shorter T_M value and the rapidly decaying ESEEM (and DEER) echo signals as nitroxide is known to form hydrogen bonds with water. (T_M represents spin phase memory time.) In the presence of largely crystallized ice, the hyperfine coupling would become much stronger due to the formation of the hydrogen bonding to nitroxide. In such a case, electron nuclear double resonance (ENDOR) experiment is a much better choice than ESEEM as ENDOR is proven to be sensitive to strong hyperfine couplings.[40] This observation adds evidence to support the non-crystalline nature of the surface water and is consistent with earlier reports from neutron scattering and NMR experiments [2], [41].

Moreover, the ESEEM data is also informative for identifying the hydrogen-bonded versus non-bonded water molecules to the N-O group of the spin label in a partially frozen state. Previous studies [38] by DFT calculations and experiments showed that water molecules that participated in a moderate hydrogen-bonding to the N-O radicals (i.e. a condition of partially crystalline state) resulted in a broad (>±2MHz in half-height width) spectral component in FT-ESEEM data, while the non-bonded water molecules gave rise to a narrow (<±0.5MHz) spectral component centered at Larmor frequency of hydrogen (or deuterium). The two spectral components would coexist in a partially frozen state around a nitroxide. [38] It is evidently shown (Figures 3b and 4b) that all of the FT-ESEEM spectra is characterized by one single narrow component, suggesting the spin label is surrounded by the water molecules that are in van der Walls contact with the nitroxide spin but are not in the correct orientations for hydrogen bonding; the spectra would otherwise become much broader. No broad component was found as the inter-pulse delay was decreased from 400 to 100 ns to probe the fast-decaying component in our study. This observation suggests the surface water remains non-crystalline/amorphous in the nanochannels (also in the vitrified bulk solvent) at 50 K. This also adds evidence to the aforementioned result concerning the non-crystalline surface water.

Long-Range Water Accessibility for the Nitroxide Spin by Pulsed ESE

Quantification of water accessibility at a somewhat larger distance range (ca. 2 nm from the nitroxide spin, which would cover several water layers) can be obtained from the analysis of the difference in the $1/T_M$ values of the ESE measurements between non-deuterated and deuterated water molecules. This difference provides a reliable estimation for the concentration of exchangeable protons, C_{ex}, within the range of ~2 nm from a nitroxide spin. The concentration can be yielded by Eq. (2), [20]

$$C_{ex} = \Delta\left(\frac{1}{T_M}\right) \times \frac{4\pi h}{0.37\mu_0(g\mu_B)^{1/2}(g_n\mu_n)^{3/2}(I(I+1))^{1/4}}$$

(2)

Where I is the nuclear spin quantum number, μ_n the nuclear magneton, μ_B the Bohr magneton, g_n and g the nuclear proton and electron g values, respectively. Figure 5 shows the analyses of extracting T_M from the ESE measurements of the two peptides, n3-s and PPm3-s. See Table 1 for the obtained T_M values (in a unit of ns) as well as the C_{ex} values (cf. Eq. 2). Note that for the bulk solvent studies of PPm3-s, the C_{ex} extracted from the data of sol(dg)/D_2O versus sol(g)/H_2O is two times greater than those for sol(s)/D_2O versus sol(s)/H_2O. This is because that the contribution from the sucrose is absent in the C_{ex} analysis since the deuterated sucrose in not conveniently available to the present study. Thus, we regard C_{ex} = 23.0 (i.e., the one extracted from the PPm3-s in sol (dg)/D_2O versus sol(g)/H_2O) as a genuine value representing the local solvation of the both peptides in the bulk vitrified solvent. The C_{ex} value for the bulk solvent is slightly less than that of the SBA15a. As compared to a proton concentration in pure water of 67 nm^{-3} (i.e., 1 g/cm^3), the obtained values are approximately half and are reasonable, since the spin-label site on the peptides should only experience fewer than a half number of the water molecules surrounding the peptide. The only exceptions are the results of the n3-s and PPm3-s in SBA15b (C_{ex} = 42.3 ~ 60.9), indicating that the two peptides in this case experience higher water densities. (Such a high C_{ex} value for a spin-labeled peptide/protein in a bulk solvent has never been reported.) These ESE results indicate that the density of exchangeable protons within the range of ~2 nm from a nitroxide spin is ordered as follows: SBA15b>SBA15a>bulk solvent. In particular, the results of the nanochannels indicate that the proton density increases substantially with the decrease in the nanochannel pore size. It is appropriate to briefly comment on the relevance of this result to other studies. Previous computational studies [42], [43] have shown that the water density is lower within the center of the nanochannel than close to the inner surface of the nanochannel. These studies, along with other experimental works, evoke a reasonable explanation for the higher water density in SBA15b than in SBA15a observed in the present study, as the nitroxide spin-hosting surface of the n3-s and PPm3-s peptides and the inner surface of the nanochannel come closer to each other in SBA15b than in SBA15a. Further discussion is presented below.

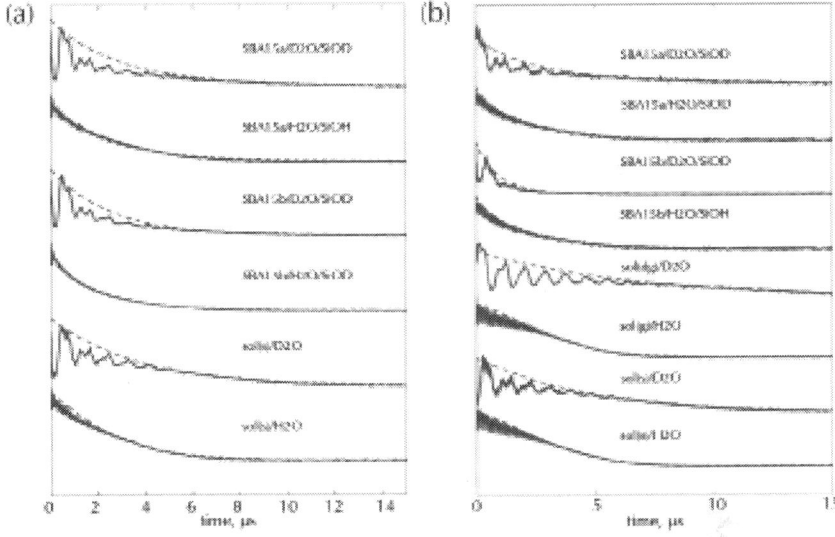

Figure 5: Long-range water accessibility study by ESE. The theoretical fits (red lines) to the ESE experimental data (blue lines) using a stretched exponential function (see Table 1) as previously described. [11] The results for the n3 -s and PPm3-s are shown in (a) and (b), respectively. The decay signals acquired by the ESE experiments were fitted over the maxima of the deuterium modulation as described in Zecevic et al. [32] to minimize the influence from destructive interference of nuclear modulations. The obtained values of the T_M (in ns) and stretching exponent x are shown in Table 1. The T_M values can be directly used to yield the surrounding proton density (C_{ex}; cf. Eq. 2) within the range of ~2 nm from a nitroxide spin.

DISCUSSION

An Intrinsic Property of the n3 peptides

The 14-mer-long polyproline model peptide was previously demonstrated by DEER (50 K) that it remained in a PPII conformation in the nanochannels and in the vitrified bulk solvent (sol(s)/H_2O), and that the P(r) of the polyproline model peptide results showed no evidence of peculiar peaks. [23] Moreover, the corresponding cw-ESR spectra of the polyproline model peptides showed no sign of large spectral broadening as this study reported for the n3 peptides (cf. Figures 1c and 2c). These DEER and cw-ESR experiments were measured in the same solvent/experimental conditions except the peptide types (PPm3 variants versus n3). Therefore, we think that it is an intrinsic property of the n3 peptide that tends to being spatially clustered (to a moderate extent) in the studied bulk solvent. This observation enhances the conclusion made for the results of n3 -d and n3 -d in the present study, providing an explanation for the peculiar minor peaks observed in Figures 1(b) and 2(b).

Close Relevance of the Cryogenic Studies to the Conditions at Room Temperature

Before delving into the implications of the experimental data, we discuss the relevance of the findings at 50 K to our understanding toward the room-temperature phenomena, which are also concerned. It is generally presumed in the protein crystallography community that room-temperature conformations of proteins, along with the structures of their surrounding water, can be preserved even at cryogenic temperatures as long as the protein crystals are cooled fast enough, as rapid cooling would leave the molecules no time to move around or wiggle considerably. Nevertheless, the regular practices in cryo-cooling protein crystals hardly, if not impossibly, attain the optimal cooling rates, and the structures observed at cryogenic temperatures turn out to be a snapshot of the structures at much lower temperatures, even down to 200 K, [44] therefore casting doubt on the biological relevance of the observations made at cryogenic temperatures. The limits on achieving the optimal cooling rates may plausibly result from the sheer sizes of typical protein crystals and the cooling method. With the common crystal size of 0.2×0.2×0.2 mm and the ordinary flash-cooling method (placing crystals directly under cold gas stream), cooling to 100 K may need the time scale of 10^{-1} seconds to reach thermal equilibrium. [45] Fortunately, our peptide-trapped-in-nanochannel samples are in the form of powder, which is equivalent to a collection of very tiny crystals, in the length scales orders of magnitude smaller than those of typical protein crystals. Due to the much-reduced sizes of the crystallites, combined with the more efficient plunge-cooling method employed here, the cooling rate for our samples are expected to be orders of magnitude higher. [45] This study thus expects that the peptide conformation and the water molecular distribution observed in the nanochannel studies faithfully reflect the conditions at room temperature. Indeed, the P(r) obtained from the DEER measurements at 50 K agrees with the NMR results collected by Kozin et al. [21] at room temperatures. Even in the bulk solution study, where the sample volumes are larger, the concerned aspects of our experimental results (i.e., backbone structure of the peptide and the interfacial water density) should still be relevant. Comparisons between the protein structures solved at room temperature and at cryogenic temperatures have demonstrated minimal cooling-induced rms deviations, within 0.08 nm, in the protein backbones. [46], [47] A careful ESR study conducted very recently has demonstrated that cooling rate, or even the very act of cooling itself, may have no observable effect on protein structural properties, manifested as the main P(r) peak positions for four different pairs of spin-labeled residues of T4 lysozyme in bulk solutions. [48] Based on the rationales reasoned above, we have considerable confidence in the room-temperature relevance of our data, even though they were collected at 50 K. This allows us to compare the experimental results with room-temperature observations from the literature and make contributions to the understanding of the room-temperature phenomena involving water.

Density Heterogeneity of the Confined Water Surrounding the Peptides

The present study has reported the ESE-proven correlation between the nanochannel diameter and the water accessibility within 2 nm from the nitroxide spin. Here, we shall discuss its physical origin and implications. The P(r) results show that the intra-molecular interspin distance of the n3 -d peptide is ~2.2 nm, regardless of the nanochannel size (Figure 1b). Given this distance and the relative positions of the two spin-labeled sites (Figure 1a), it is reasonable to expect an equivalent diameter of >2.2 nm for freely rotating n3 peptide. Assuming that diameter is ~2.5 nm and the peptide is located near the nanochannel center, the distance between the peptide surface and the channel wall would be ~2.55 nm for SBA15a (with the diameter of 7.6 nm) and ~1.8 nm for SBA15b (with the diameter of 6.1 nm). Since the nitroxide spin is ~0.7 nm from the n3 peptide surface, the range covered by the ESE measurements in assessing the water accessibility should be ~2.7 nm from the peptide surface, encompassing all the relevant space between the peptide and the channel wall for both of the SBA15a and SBA15b cases. In other words, the density of the exchangeable protons (presumably from water molecules) extracted from the ESE measurements is in fact the water density averaged over the entire inter-surface space. It has long been established from experimental and computational studies that water in the first hydration layer (~0.3 nm thick) of a surface exhibits a density distinctly greater than that of bulk water, due to the perturbations on water by that surface. [49]–[51] Since this surface effect is always present and independent of the nanochannel diameter so long as the inter-surface space does not shrink to a few water layer thick, the amount of the bulk-like water halfway between the two surfaces would gradually diminish when the nanochannel diameter keeps decreasing, allowing the surface water to become dominant and the averaged water density to rise. This explains why the averaged water density within 2.7 nm from the peptide surface is greater in SBA15b than in SBA15a. Furthermore, given the extent of the density elevation (at least about two times) upon adopting SBA15b (see the previous section) and the fact that water accessibility within 0.35 nm from the nitroxide spin is of the same magnitude between SBA15a and SBA15b (see the results from the ESEEM measurements; Table 1), we argue that water layers beyond the first hydration water also exhibit higher-than-bulk-water densities and contribute to the observation in the ESE measurements. Since the inter-surface space for SBA15b is ~1.8 nm thick if the peptide exhibits an equivalent diameter of ~2.5 nm (see above), we speculate that the peptide surface and the nanochannel wall may each perturb up to 3 layers or 0.9 nm thick of water from the surfaces (i.e., $2\times3\times0.3 = 1.8$ nm), leading to the higher-than-bulk-water densities in these layers. Indeed, earlier experimental and computational studies indicate that the influence of the surface effect is well beyond the first hydration layer [52]–[54] and probably up to 1 nm from the surface, as indicated by a femtosecond-resolved fluorescence study [53]; Kim et al. [54] even went as far as classifying the 3 surface-perturbed water layers as one layer of "well-ordered rigid water" and two layers of "quasi-bound water". Our study reports one of the few experimental, albeit indirect, evidence in the context of water density to support the existence of multiple (more than one) surface-perturbed water layers, and may provide a foundation to investigate the structural origin of the "dynamic" hydration layer proposed by Zewail's and Zhang's groups [53]–[55].

Support to the above speculation can also be gained by examining the ESE results of the PPm3-s. Previously it was shown that this long helical polyproline-based peptide remains structurally intact and undergoes a large degree of rotational anisotropy and orientational ordering inside the SBA15 nanochannels. [23] The PPm3 was demonstrated to align favorably with its long axis being parallel to the longitudinal axis of the nanochannels. The present study reports that the C_{ex} values are 60.9 and 28.6 for SBA15b and SBA15a, respectively, showing a good agreement with the finding for the n3 -s. Both the peptides, though in two different secondary structures, report a similar trend for the $C_{ex'}$ suggesting the water density variation along the transverse direction of the nanochannel is the key to the observed changes in the ESE measurements and could be independent of the topological details of a surface. Besides, the present study verifies the density heterogeneity along the peptide surface normal in the nanochannels.

The results from the ESEEM measurements tell another interesting story. The ESEEM study reports the water accessibility within the van-der-Waals contact of 0.35 nm from the nitroxide spin; this range corresponds to the first hydration water of the peptides. The values for the two peptides in the nanochannels with different diameters are roughly the same and within the range for a typical solvent-exposed site on a protein. This indicates that the first hydration water accessibility is essentially unchanged (within errors) in nanospace, irrespective of the peptide secondary structure, peptide length, or the nanochannel diameter. This fact further enhances our argument in the previous two paragraphs that the surface effect is unaffected by the nanochannel size. In a comparison of the bulk and nanochannel studies, the results of the analysis are similar if sucrose is used. However, the value for the bulk solution becomes double if deuterated glycerol is used. As shown in Eq. (2), the value is not affected by the non-deuterated cryoprotectants. Therefore, this ESEEM study manifests that the interfacial hydration accessibility (i.e., the first layer) for the bulk solvents is approximately two times greater than that in the nanochannels, while the secondary structure remains approximately unchanged. This suggests that variations in the interfacial solvation density are not of considerable influence to the stability of the peptide structures. The present finding could support the concept that presence of the first hydration layer, in contrast to the long-held belief, is a sufficient but not indispensible condition for ensuring the proper function and structure of a protein, as demonstrated in two previous studies that i) a completely unsolvated polyalanine helix in the gas phase could remain up to at least 723 K; [7] ii) water may even be replaced with a polymer in rendering a protein the dynamical flexibility and biological functionality. [56] Taken together the present and the previous studies, we argue that the first hydration layers might play a role in loosening and providing flexibility to the structure of a protein rather than stabilizing it.

Altogether, this appears to indicate that the first hydration water accessibility is independent of the surface topography of the peptide; the spatial arrangement of the water molecules in the first hydration layer has no considerable impact on the peptide conformation. Moreover, the first hydration water accessibility seems to be unaffected by the condition of the bulk water, because this accessibility (values) changes little as compared to the SBA15a and SBA15b, where the bulk water for the latter is considerably diminished or even absent. All these observations evoke the need of a picture alternative to that involving the bulk water effect on hydration water, as proposed by Merzel and Smith [50], to explain how the surface topography affects the first hydration water density. How much the surface topography affects the first hydration water density is even a matter of debate in lieu of our ESEEM data.

Inter-peptide Interactions are excluded from the Analyses

This study shows that given the same [n3 -d] concentration, the baseline of the DEER measurement for the SBA15b decays rapidly as compared to the baselines of the SBA15a and bulk solution studies (Figure 1a). This indicates the bulk spin concentration (denoted as [n3 -d]' hereafter) is increased upon the encapsulation into the SBA15b, suggesting the spatial distribution of the n3 -d peptides is adjusted for the nanochannel structures. The apparent spin concentration, i.e., [n3 -d]', in the SBA15b, if too high, would likely give rise to a greater degree of the unwanted inter-spin interactions (e.g., instantaneous diffusion, which becomes important as spin concentration is sufficiently high,) and the n3-d peptide clustering/aggregation in the nanochannels. However, the following summary of the results indicates that the two suspected conditions hardly exist in this study. Cw-ESR lineshape is sensitive to inter-spin distances less than 2 nm. [57] The cw-ESR spectra (Fig. 1c) of the SBA15a and SBA15b are similar to each other and show no sign of the inter-peptide dipolar broadening. (Same observation can be obtained from the n3-d study shown in Figure 2c.) This indicates that the inter-peptide distances in the nanochannels are long enough (>2 nm) so that the inter-molecular dipolar interactions are too weak to appear in the cw-ESR spectra. To investigate whether or not the instantaneous diffusion plays an important role in the nanochannel study, we thus performed the DEER measurement with increasing the length of the observer $\varpi/2$ and ϖ pulses from 16 and 32 ns, respectively, to 32 and 64 ns, respectively. If the instantaneous diffusion plays an important role in the nanochannel studies, we would observe an increase in the signal-to-noise ratio (SNR) of the DEER data with the increased pulse length. This is because of that as the instantaneous diffusion is important, excitation of a larger fraction of spins may not necessarily translate to higher sensitivity. [58] Our measurement found that no increase in the SNR was observed in the experiment. Taken together, the above observations rule out the two suspected conditions. The individual peptides are sufficiently separated (>2 nm) in both the SBA15a and SBA15b nanochannels. The observed increase in the proton density for SBA15b is indeed attributable to the increased solvent (proton) density that results from the decrease in the nanochannel pore size. This study strongly suggests that the analysis results of the ESEEM and ESE measurements for the nanochannel studies reflect largely the difference in the solvation shells (i.e., the interactions between the nitroxide spin and the nearby solvent molecules,) rather than the inter-peptide interactions.

Last, we comment on the effect of the increased [n3-d]' on the P(r) determination of the SBA15b. Previously, the n3-d was studied by the DEER technique at 50 K, with [n3-d] = 1.5 mM (which is three times greater as compared to the present study) in the nanochannel of the MSU material. [11] The P(r) obtained in the previous study is closely similar to the distance distributions of the SBA15a and SBA15b in the present study. This observation indicates that the variation of the bulk spin concentration affects insignificantly the determination of the distance distribution by Tikhonov regularization method. The use of nanochannel is useful to avoid clustering of spin-labeled molecules, making it possible to perform DEER in higher concentrations.

CONCLUSIONS

In summary, we have studied two different spin-labeled peptides in the (mostly) vitrified bulk solvents and in the nanochannels of several different pore diameters (6.1 ~ 7.6 nm) using ESR techniques (cw, DEER, ESEEM, and ESE) at 50 K. The studied peptides include a 26-residue-long polypeptide in two conformations (-helix versus -hairpin), inducible by the change of the solvent, and a 14-residue-long polyproline model peptide. This study demonstrates the peptide being confined but structurally unaltered in the nanospace, with its surrounding water exhibiting density heterogeneity along the peptide surface normal. The DEER measurements demonstrate that the peptide conformations remain approximately unchanged upon the encapsulation into the nanochannels. The ESEEM results show that the water accessibility in the immediate vicinity of the nitroxide label (the first hydration layer, within the van der Waals contacts, ~0.35 nm) at the water-peptide interface changes little with the nanochannel dimensions but reduce by half as comparing to the accessibility obtained for the bulk solvent studies. Nevertheless, the ESE results show that the hydration density in the range of ~2 nm from the nitroxide spin (or ~2.7 nm from the peptide surface) varies notably and increases with the decreasing nanochannel diameter; this is considered to be a consequence of diminishing or completely removal of the bulk-like water when the nanochannel size is reduced. Indeed, the ESE shows the long-range water accessibility in the SBA15a nanochannel (i.e., of the largest pore size in this study) changes little as compared with the bulk solvent. Our conclusions are based on the observed relative changes of water density among different confinement conditions at a constant temperature of 50 K; therefore, the changes upon temperature variations (e.g., abrupt changes of water properties accompanying phase transition upon cooling) are irrelevant to the validity of our conclusion. We infer from these observations that i) the water accessibility of the first hydration layer appears to be essentially unchanged, regardless of the water properties outside the first hydration layer or the surface topological details of a peptide; ii) perturbations on water, induced by the peptide and nanochannel inner surfaces, may each affect up to several (2~3) layers of water from these surfaces, and is independent of the surface topological details, too. This second conclusion experimentally demonstrates the range of the surface effect on the surrounding water, while the first conclusion might evoke the need of alternative insight to explain why water exhibits density variation across the protein surface to the bulk. Moreover, this study shows the polypeptides are confined, but free and hydrated in nanochannels. The secondary structures of the peptides appear to be insensitive to the solvation properties such as the solvation accessibility, as this study (which is solely based on the studies of the n3 and PPm3 peptide variants) demonstrates the peptides retain their structures even the solvation accessibility is changed by a large extent. This finding suggests that the density property of the interfacial solvation (namely, the solvation in the immediate vicinity of the spin-labeled sites) is not essential to the stability of a peptide structure. This study has opened up exciting avenues for looking into the protein-water interactions and the biomolecules in nano-confinements using ESR with nanochannels.

ACKNOWLEDGMENTS

The authors are grateful for the mesoporous materials provided by Prof. Chia-Min Yang. All of the cw/pulsed ESR measurements were carried out in NSC Research Instrument Center of Taiwan located at NTHU.

AUTHOR CONTRIBUTIONS

Conceived and designed the experiments: YCL YWC. Performed the experiments: YCL YWC. Analyzed the data: YFC YWC. Contributed reagents/materials/analysis tools: YWC. Wrote the paper: YFC YWC.

REFERENCES

1. Nucci NV, Pometun MS, Wand AJ (2011) Mapping the hydration dynamics of ubiquitin. J Am Chem Soc 133: 12326–12329. doi:

10.1021/ja202033k

2. Chen SH, Liu L, Fratini E, Baglioni P, Faraone A, et al. (2006) Observation of fragile-to-strong dynamic crossover in protein hydration water. Proc Natl Acad Sci USA 103: 9012–9016. doi: 10.1073/pnas.0602474103

3. Halle B (2004) Protein hydration dynamics in solution: a critical survey. Philos T Roy Soc B 359: 1207–1223. doi: 10.1098/rstb.2004.1499

4. Swenson J, Jansson H, Bergman R (2006) Relaxation processes in supercooled confined water and implications for protein dynamics. Phys Rev Lett 96: 247802. doi: 10.1103/physrevlett.96.247802

5. Rasmussen BF, Stock AM, Ringe D, Petsko GA (1992) Crystalline Ribonuclease-a Loses Function Below the Dynamic Transition at 220-K. Nature 357: 423–424. doi: 10.1038/357423a0

6. Schiro G, Caronna C, Natali F, Koza MM, Cupane A (2011) The "Protein Dynamical Transition" Does Not Require the Protein Polypeptide Chain. J Phys Chem Lett 2: 2275–2279. doi: 10.1021/jz200797g

7. Kohtani M, Jones TC, Schneider JE, Jarrold MF (2004) Extreme stability of an unsolvated alpha-helix. J Am Chem Soc 126: 7420–7421. doi: 10.1021/ja048766c

8. Ortony JH, Cheng CY, Franck JM, Kausik R, Pavlova A, et al. (2011) Probing the hydration water diffusion of macromolecular surfaces and interfaces. New J Phys 13: 015006. doi: 10.1088/1367-2630/13/1/015006

9. Banerjee D, Bhat SN, Bhat SV, Leporini D (2012) Molecular Probe Dynamics Reveals Suppression of Ice-Like Regions in Strongly Confined Supercooled Water. Plos One 7.

10. Banerjee D, Bhat SN, Bhat SV, Leporini D (2009) ESR evidence for 2 coexisting liquid phases in deeply supercooled bulk water. P Natl Acad Sci USA 106: 11448–11453. doi: 10.1073/pnas.0900734106

11. Huang YW, Lai YC, Tsai CJ, Chiang YW (2011) Mesopores provide an amorphous state suitable for studying biomolecular structures at cryogenic temperatures. Proc Natl Acad Sci USA 108: 14145–14150. doi: 10.1073/pnas.1102395108

12. Huang YW, Chiang YW (2011) Spin-label ESR with nanochannels to improve the study of backbone dynamics and structural conformations of polypeptides. Phys Chem Chem Phys 13: 17521–17531. doi: 10.1039/c1cp20986h
13. Smith JV, Arnold FP, Parsons I, Lee MR (1999) Biochemical evolution III: Polymerization on organophilic silica-rich surfaces, crystal-chemical modeling, formation of first cells, and geological clues. Proc Natl Acad Sci USA 96: 3479–3485. doi: 10.1073/pnas.96.7.3479
14. Fleissner MR, Bridges MD, Brooks EK, Cascio D, Kalai T, et al. (2011) Structure and dynamics of a conformationally constrained nitroxide side chain and applications in EPR spectroscopy. Proc Natl Acad Sci USA 108: 16241–16246. doi: 10.1073/pnas.1111420108
15. Tong JS, Borbat PP, Freed JH, Shin YK (2009) A scissors mechanism for stimulation of SNARE-mediated lipid mixing by cholesterol. Proc Natl Acad Sci USA 106: 5141–5146. doi: 10.1073/pnas.0813138106
16. Park SY, Borbat PP, Gonzalez-Bonet G, Bhatnagar J, Pollard AM, et al. (2006) Reconstruction of the chemotaxis receptor-kinase assembly. Nat Struct Mol Biol 13: 400–407. doi: 10.1038/nsmb1085
17. Pannier M, Veit S, Godt A, Jeschke G, Spiess HW (2000) Dead-Time Free Measurement of Dipole-Dipole Interactions between Electron Spins. J Magn Reson 142: 331–340. doi: 10.1006/jmre.1999.1944
18. Milov AD, Maryasov AG, Tsvetkov YD, Raap J (1999) Pulsed ELDOR in spin-labeled polypeptides. Chem Phys Lett 303: 135–143. doi: 10.1016/s0009-2614(99)00220-1
19. Carmieli R, Papo N, Zimmermann H, Potapov A, Shai Y, et al. (2006) Utilizing ESEEM spectroscopy to locate the position of specific regions of membrane-active peptides within model membranes. Biophys J 90: 492–505. doi: 10.1529/biophysj.105.062992
20. Volkov A, Dockter C, Bund T, Paulsen H, Jeschke G (2009) Pulsed EPR Determination of Water Accessibility to Spin-Labeled Amino Acid Residues in LHCIIb. Biophys J 96: 1124–1141. doi: 10.1016/j.bpj.2008.09.047

21. Kozin SA, Bertho G, Mazur AK, Rabesona H, Girault JP, et al. (2001) Sheep prion protein synthetic peptide spanning helix 1 and beta-strand 2 (residues 142–166) shows beta-hairpin structure in solution. J Biol Chem 276: 46364–46370. doi: 10.1074/jbc.m108014200

22. Megy S, Bertho G, Kozin SA, Debey P, Hoa GHB, et al. (2004) Possible role of region 152–156 in the structural duality of a peptide fragment from sheep prion protein. Protein Sci 13: 3151–3160. doi: 10.1110/ps.04745004

23. Tsai CJ, Chiang YW (2012) Effects of Anisotropic Nanoconfinement on Rotational Dynamics of Biomolecules: An Electron Spin Resonance Study. J Phys Chem C 116: 19798–19806. doi: 10.1021/jp306478c

24. Yang CM, Zibrowius B, Schmidt W, Schuth F (2004) Stepwise removal of the copolymer template from mesopores and micropores in SBA-15. Chem Mater 16: 2918–2925. doi: 10.1021/cm049526z

25. Sung TC, Chiang YW (2010) Identification of complex dynamic modes on prion protein peptides using multifrequency ESR with mesoporous materials. Phys Chem Chem Phys 12: 13117–13125. doi: 10.1039/c0cp00685h

26. Morrow BA, Mcfarlan AJ (1994) Infrared Study of Chemical and H-D Exchange Probes for Silica Surfaces. Adv Chem Ser 234: 183–198. doi: 10.1021/ba-1994-0234.ch009

27. Chiang YW, Borbat PP, Freed JH (2005) The Determination of Pair Distance Distributions by Pulsed ESR Using Tikhonov Regularization. J Magn Reson 172: 279–295. doi: 10.1016/j.jmr.2004.10.012

28. Chiang YW, Borbat PP, Freed JH (2005) Maximum entropy: A complement to Tikhonov regularization for determination of pair distance distributions by pulsed ESR. J Magn Reson 177: 184–196. doi: 10.1016/j.jmr.2005.07.021

29. Jeschke G (2012) DEER Distance Measurements on Proteins. Annual Review of Physical Chemistry, Vol 63 63: 419–446. doi: 10.1146/annurev-physchem-032511-143716

30. Ichikawa T, Kevan L, Bowman MK, Dikanov SA, Tsvetkov YD (1979) Ratio Analysis of Electron-Spin Echo Modulation

Envelopes in Disordered Matrices and Application to the Structure of Solvated Electrons in 2-Methyltetrahydrofuran Glass. J Chem Phys 71: 1167–1174. doi: 10.1063/1.438462
31. Stoll S, Britt RD (2009) General and efficient simulation of pulse EPR spectra. Phys Chem Chem Phys 11: 6614–6625. doi: 10.1039/b907277b
32. Zecevic A, Eaton GR, Eaton SS, Lindgren M (1998) Dephasing of electron spin echoes for nitroxyl radicals in glassy solvents by non-methyl and methyl protons. Mol Phys 95: 1255–1263. doi: 10.1080/00268979809483256
33. Parkin S, Rupp B, Hope H (1996) Structure of bovine pancreatic trypsin inhibitor at 125 K: Definition of carboxyl-terminal residues Gly57 and Ala58. Acta Crystallogr D 52: 18–29. doi: 10.1107/s0907444995008675
34. Bourgeois D, Royant A (2005) Advances in kinetic protein crystallography. Curr Opin Struc Biol 15: 538–547. doi: 10.1016/j.sbi.2005.08.002
35. Theisen MJ, Misra I, Saadat D, Campobasso N, Miziorko HM, et al. (2004) 3-hydroxy-3-methylglutaryl-CoA synthase intermediate complex observed in "real-time". P Natl Acad Sci USA 101: 16442–16447. doi: 10.1073/pnas.0405809101
36. Borbat PP, Freed JH (2007) Measuring distances by pulsed dipolar ESR spectroscopy: Spin-labeled histidine kinases. Methods Enzymol 423: 52–116. doi: 10.1016/s0076-6879(07)23003-4
37. von Hagens T, Polyhach Y, Sajid M, Godt A, Jeschke G (2013) Suppression of ghost distances in multiple-spin double electron-electron resonance. Phys Chem Chem Phys 15: 5854–5866. doi: 10.1039/c3cp44462g
38. Erilov DA, Bartucci R, Guzzi R, Shubin AA, Maryasov AG, et al. (2005) Water concentration profiles in membranes measured by ESEEM of spin-labeled lipids. J Phys Chem B 109: 12003–12013. doi: 10.1021/jp050886z
39. Deligiannakis Y, Louloudi M, Hadjiliadis N (2000) Electron spin echo envelope modulation (ESEEM) spectroscopy as a tool to investigate the coordination environment of metal centers. Coordin Chem Rev 204: 1–112. doi: 10.1002/chin.200037298
40. Smirnova TI, Smirnov AI, Paschenko SV, Poluektov OG (2007)

Geometry of hydrogen bonds formed by lipid bilayer nitroxide probes: A high-frequency pulsed ENDOR/EPR study. J Am Chem Soc 129: 3476–3477. doi: 10.1021/ja068395v

41. Kuntz ID, Brassfie.Ts, Law GD, Purcell GV (1969) Hydration of Macromolecules. Science 163: 1329–1331. doi: 10.1126/science.163.3873.1329

42. Milischuk AA, Ladanyi BM (2011) Structure and dynamics of water confined in silica nanopores. J Chem Phys 135: 174709. doi: 10.1063/1.4733970

43. Gallo P, Rovere M, Chen SH (2010) Dynamic Crossover in Supercooled Confined Water: Understanding Bulk Properties through Confinement. J Phys Chem Lett 1: 729–733. doi: 10.1021/jz9003125

44. Halle B (2004) Biomolecular cryocrystallography: Structural changes during flash-cooling. P Natl Acad Sci USA 101: 4793–4798. doi: 10.1073/pnas.0308315101

45. Kriminski S, Kazmierczak M, Thorne RE (2003) Heat transfer from protein crystals: implications for flash-cooling and X-ray beam heating. Acta Crystallogr D 59: 697–708. doi: 10.1107/s0907444903002713

46. Juers DH, Matthews BW (2001) Reversible lattice repacking illustrates the temperature dependence of macromolecular interactions. J Mol Biol 311: 851–862. doi: 10.1006/jmbi.2001.4891

47. Sandalova T, Schneider G, Kack H, Lindqvist Y (1999) Structure of dethiobiotin synthetase at 0.97 angstrom resolution. Acta Crystallogr D 55: 610–624. doi: 10.1107/s090744499801381x

48. Georgieva ER, Roy AS, Grigoryants VM, Borbat PP, Earle KA, et al. (2012) Effect of freezing conditions on distances and their distributions derived from Double Electron Electron Resonance (DEER): A study of doubly-spin-labeled T4 lysozyme. J Magn Reson 216: 69–77. doi: 10.1016/j.jmr.2012.01.004

49. Engemann S, Reichert H, Dosch H, Bilgram J, Honkimaki V, et al.. (2004) Interfacial melting of ice in contact with SiO2. Phys Rev Lett 92.

50. Merzel F, Smith JC (2002) Is the first hydration shell of lysozyme of higher density than bulk water? P Natl Acad Sci USA 99: 5378–5383. doi: 10.1073/pnas.082335099
51. Svergun DI, Richard S, Koch MHJ, Sayers Z, Kuprin S, et al. (1998) Protein hydration in solution: Experimental observation by x-ray and neutron scattering. P Natl Acad Sci USA 95: 2267–2272. doi: 10.1073/pnas.95.5.2267
52. Pinnick ER, Erramilli S, Wang F (2010) Computational investigation of lipid hydration water of L-alpha 1-palmitoyl-2-oleoyl-sn-glycero-3-phosph ocholineat three hydration levels. Mol Phys 108: 2027–2036. doi: 10.1080/00268976.2010.503199
53. Zhang LY, Yang Y, Kao YT, Wang LJ, Zhong DP (2009) Protein Hydration Dynamics and Molecular Mechanism of Coupled Water-Protein Fluctuations. J Am Chem Soc 131: 10677–10691. doi: 10.1021/ja902918p
54. Kim J, Lu WY, Qiu WH, Wang LJ, Caffrey M, et al. (2006) Ultrafast hydration dynamics in the lipidic cubic phase: Discrete water structures in nanochannels. J Phys Chem B 110: 21994–22000. doi: 10.1021/jp062806c
55. Qiu WH, Kao YT, Zhang LY, Yang Y, Wang LJ, et al. (2006) Protein surface hydration mapped by site-specific mutations. P Natl Acad Sci USA 103: 13979–13984. doi: 10.1073/pnas.0606235103
56. Gallat FX, Brogan APS, Fichou Y, McGrath N, Moulin M, et al. (2012) A Polymer Surfactant Corona Dynamically Replaces Water in Solvent-Free Protein Liquids and Ensures Macromolecular Flexibility and Activity. J Am Chem Soc 134: 13168–13171. doi: 10.1021/ja303894g
57. Altenbach C, Oh KJ, Trabanino RJ, Hideg K, Hubbell WL (2001) Estimation of inter-residue distances in spin labeled proteins at physiological temperatures: Experimental strategies and practical limitations. Biochemistry 40: 15471–15482. doi: 10.1021/bi011544w
58. Jeschke G, Polyhach Y (2007) Distance measurements on spin-labelled biomacromolecules by pulsed electron paramagnetic resonance. Phys Chem Chem Phys 9: 1895–1910. doi: 10.1039/b614920k

Chapter 8

Removal Efficiency of Radioactive Cesium and Iodine Ions by a Flow-Type Apparatus Designed for Electrochemically Reduced Water Production

Takeki Hamasaki, Noboru Nakamichi, Kiichiro Teruya, and Sanetaka Shirahata

Department of Bioscience and Biotechnology, Faculty of Agriculture, Kyushu University, Higashi-ku, Fukuoka, Japan

ABSTRACT

The Fukushima Daiichi Nuclear Power Plant accident on March 11, 2011 attracted people's attention, with anxiety over possible radiation hazards. Immediate and long-term concerns are around protection from external and internal exposure by the liberated radionuclides. In particular, residents living in the affected regions are most concerned about ingesting contaminated foodstuffs, including drinking water. Efficient removal of radionuclides from rainwater and drinking water has been reported using several pot-type filtration devices. A currently used flow-type test apparatus is expected to simultaneously provide radionuclide elimination prior to ingestion and protection from internal exposure by accidental ingestion of radionuclides through the use of a micro-carbon carboxymethyl cartridge unit and an electrochemically reduced water production unit, respectively. However, the removability of radionuclides from contaminated tap water has not been tested to date. Thus, the current research was undertaken to assess the capability of the apparatus to remove radionuclides from artificially contaminated tap water. The results presented here demonstrate that the apparatus can reduce radioactivity levels to below the detection limit in applied tap water containing either 300 Bq/kg of ^{137}Cs or 150 Bq/kg of ^{125}I. The apparatus had a removal efficiency of over 90% for all concentration ranges of radio–cesium and –iodine tested. The results showing efficient radionuclide removability, together with previous studies on molecular hydrogen and platinum nanoparticles as reactive oxygen species scavengers, strongly suggest that the test apparatus has the potential to offer maximum safety against radionuclide-contaminated foodstuffs, including drinking water.

INTRODUCTION

The Great East Japan Earthquake of magnitude 9 struck the northeastern coast of Japan on March 11, 2011. The earthquake caused a catastrophic tsunami, with the wave height of nearly 40.5 m, which caused failures in the nuclear reactor cooling system in the Fukushima Daiichi Nuclear Power Plant (FDNPP) [1], [2]. Soon after, these failures triggered hydrogen explosions in the nuclear reactors, discharging radioactive steam and liberating various radionuclides into the air over several days [2], [3]. Following the incident, natural factors such as wind flow, air streams, and rainfall caused dispersion and precipitation of various levels of radionuclides on land surfaces and vegetation in the Tohoku and Kanto regions [4]–[8]. Radionuclides were also detected in Fukuoka, 1,000 km away from the FDNPP [9], indicating the wide spread of the radioactive plume over Japan. Urgent action to cope with the situation involves decontamination of terrestrial and aquatic radioactivity sources, including drinking water. Incineration of contaminated materials such as plants, wood bark, garbage, and house wreckage is one choice for disposition, although it leaves cesium-enriched ash. An entire system for safe incineration, removal of ash radioactivity and safe disposal has been reported, with promising results [10]. Numerous conventional methods using ion exchange, various membrane processes, coagulation and co-precipitation and other technologies for eliminating radionuclides from radioactive wastewaters have been reported to be effective [11], [12]. Numerous approaches have been shown to remove radionuclides from contaminated water, including a mixture of activated carbon and/or zeolite-based media [13]–[15], co-precipitation with zinc hexacyanoferrate (II) followed by precipitation [16], sorption of radionuclides with biomaterials such as diatomite [17], Prussian blue immobilized diatomite or alginate/calcium beads or magnetic nanoparticles [18]–[20], arca shell [21], sulphuric acid-modified persimmon waste[22], nickel (II) hexacyanoferrate (III) functionalized walnut shell [23], mesoporous silica monoliths conjugated with dibenzo-18-crown-6 ether [24], and cobalt ferrocyanide impregnated anion exchange beads [25]. Additionally, a layered chalcogenide with a CdI_2 crystal structure for adsorbing several cations has been explored [26].

Although these technologies are encouraging for removal of various levels of radionuclides and further improvements are expected to arise in the future, securing safe drinking water is also of prime importance. Rainwater samples collected in Fukushima in early April, 2011 have been reported to contain ^{131}I (1470±26.5 Bq/L), ^{134}Cs (100±25.3 Bq/L) and ^{137}Cs (129±9.47 Bq/L)[27]. The fallout contaminates surface waters, including lakes and rivers, which are the main sources for preparing tap water to supply the residents in these regions. As a result, drinking water prepared from several water purification plants was reported to be contaminated. Subsidiary methods to reinforce conventional water purification systems have been reported to eliminate radioactivity from contaminated water sources. The efficacies of the coagulation-flocculation-sedimentation method in water purification plants, with removal efficiencies of 17% and 56% for ^{131}I and ^{134}Cs, respectively [28], [29], and radionuclide absorption by algal strains for environmental remediation [30], [31] have been assessed. Another significant point to consider is the contamination of drinking water via distribution system such as pipes, storage tanks, water pumps and heaters, which may be persistent contaminating sources. A recent review concluded that cesium appears to be removed by flushing water pipes with a low pH solution containing sodium or magnesium as ion competitors [32]. However, further assessment will be required before applying this approach to the vast areas of regional contamination. Approximately one month later, the radioactivity levels had decreased to below the limit values in the water purification plants [33]. Whereas even after 2 years, total Cs radioactivities above the limit values are reported in some foodstuffs, such as Chinese mushrooms, rice, soybean, adzuki-bean and several fish obtained from the areas surrounding the FDNPP [34], [35]. Moreover, low levels of radioactive Cs species are still detected in the drinking water of many cities around FDNPP [36]. These results imply that the fallout still remains on land surfaces and nearby mountain areas and that rainfall wash down is a highly probable contaminant of tap water sources [7], [8]. Precautions to avoid consumption of such foodstuffs, including drinking water, have been taken by measuring radioactivity levels prior to distribution. Nevertheless, following the accident, the concentrations of ^{131}I in the tap water distributed by these purification plants were 210 Bq/L in Tokyo, 189 Bq/L in Ibaraki, and 220 Bq/L in Chiba, all of which exceeded the upper limit of ^{131}I concentration set as 100 Bq/L for infants under 1 year of age by the Ministry of Health, Labour and Welfare, 1947 [3], [37]. Therefore, it is highly desirable to have terminal security systems that can achieve the removal of even lower levels of radioactive contaminants in tap water because, for example, radiocesium accumulates in the body. However, only limited studies examining removability of radionuclides from household water purifiers are available to date. Several domestic pot-type water purifiers have been suggested as a possible final security treatment to eliminate contaminated radionuclides in tap water [27], [38]. Although most of these pot-type water purifiers are efficacious, with varying degrees of radionuclide removal from contaminated water, they are useless against the biological effects exerted by unconscious ingestion of radionuclides via drinking water and/or foodstuffs.

Ionizing radiation emitted by ingested radionuclides causes water radiolysis by acting on the water molecules, which comprise approximately 80% of body weight [39]. Water radiolysis yields a variety of reactive oxygen species (ROS) including hydrogen peroxide (H_2O_2), the hydroxyl radical (•OH), superoxide anion radicals (•O_2^-), and other molecular species [40]. These free radicals cause extensive oxidative damage to biologically critical macromolecules such as DNA, RNA, proteins and lipids [41]–[45]. Such damage eventually induces cellular apoptosis or carcinogenic transformation [46], [47]. Therefore, an ideal apparatus should have the potential to provide both the elimination of radionuclides prior to ingestion and protection from detrimental ROS effects generated by the accidentally and/or unconsciously internalized radionuclides.

Considering these requirements, an apparatus designed to produce electrochemically reduced water (ERW) could be thought to fulfill such demands because it contains two functional units; an electrolysis unit for molecular hydrogen enrichment, and a micro-carbon carboxymethyl (CM) cartridge unit for removing various impurities. ERW produced from tap water by this apparatus contains as much as 0.587 ppm dissolved hydrogen (Table 1, [48]). Dissolved molecular hydrogen has been shown to exert a radioprotective effect in both *in vitro* and *in vivo* studies [49], [53]. These compelling results strongly support the suggestion that molecular hydrogen dissolved in ERW could function as a radioprotective agent in the body. Moreover, ERW was shown to contain platinum nanoparticles (Pt NPs) at up to 2.5 ppb as an ROS scavenger, liberated from Pt-electrodes during electrolysis [39], [54].

Table 1: Characteristics of the sample waters

EPW							
	Tap Water	Filtered Water	Lv 1	Lv 2	Lv 3	Lv 4
pH	7.6±0.0	7.6±0.0	8.0±0.0	8.5±0.0	9.1±0.0	9.4±0.1
ORP (mV)	555.3 ± 15.5	550.0 ± 20.1	140.0 ± 5.0	110.0 ± 7.5	-673.3±2.5	-688.0±9.5
EC (ms/m)	49.3±0.1	49.5±0.1	49.7±0.1	49.7±0.1	49.0±0.2	48.1 ±0.2

DH (ppb, gg/1)	N.D.	N.D.	70.0 ± 19.3	163.3 ± 18.0	321.7±47.5	587.0±44.6
DO (ppm, WI)	7.5±0.0	7.5±0.0	7.5±0.1	7.1 ±0.1	6.6±0.2	6.1 ±0.3

Filtered water: tap water was passed through the micro carbon cartridge without electrolysis. Lv 1: electrochemically reduced water (ERW) generated by electrolyzing the filtered water at level 1 with constant electric current at 50 volts (V) upper limit voltage and a flow rate of 1.8–2.0 l/min. Likewise, other ERWs were produced using identical conditions, except selecting the Lv 2 to Lv 4 switch. ORP: oxidation-reduction potential. EC: electrical conductivity. DH: dissolved hydrogen. DO: dissolved oxygen. Measurements were conducted at ambient temperatures. N.D.: Not Detected.

As for the second requirement, a micro-carbon CM cartridge unit composed of a nonwoven-fabric filter, several types of activated carbon and an ion-exchange material was present in the current test apparatus to remove particulate matters, microorganisms and 13 designated impurities [55]. However, this micro-carbon CM cartridge has not been assessed for its ability to remove radionuclides from contaminated tap water. Therefore, the present research was aimed at evaluating whether the test apparatus as a whole is capable of removing radionuclides from contaminated tap water.

MATERIALS AND METHODS

Chemicals

Cesium chloride (CsCl) and potassium iodide (KI) were purchased from Wako Pure Chemical Industries (Osaka, Japan).

Radioisotopes

[137]CsCl [0.2021 MBq/g] and Na[125]I [12.950 TBq/g] were purchased from Japan Radioisotope Association (JRIA, Tokyo, Japan). We used ^{125}I because Kyushu University Radioisotope Center has an approval

to use this radionuclide. Tap water distributed by the Fukuoka City Waterworks Bureau, Fukuoka, Japan was used in all experiments except ultrapure water (Milli Q water, Merck Millipore, Tokyo, Japan) for the preparation of standard solutions for inductively coupled plasma-mass spectrometry (ICP-MS) analysis.

Electrochemically Reduced Water (ERW)-Producing Apparatus

A water flow-type apparatus, Trim Ion NEO, was provided by Nihon Trim Co. Ltd., Osaka, Japan as the test apparatus. This test apparatus is composed of two units, a micro-carbon CM cartridge unit (Fig. 1B) and an electrolysis unit (Fig. 1C). Tap water flows into the cartridge unit, where tap water passes through the nonwoven-fabric filter to remove macroparticles, and pre-cleaned water flows into mixed layers of activated charcoal powders and cationic ion-exchange material to remove most of the impurities, including dissolved lead and 13 other elements that must be removed. The remaining contaminants, such as microorganisms and iron rust particles larger than 0.1 µm in size, are also eliminated by the cartridge (Fig. 1B). The micro-carbon CM cartridge unit is certified to withstand filtration of at least 12 tons of tap water per year or 35 liters per day for 1 year. In the present study, we used a new cartridge unit for each experiment. Purified tap water flows into the electrolysis unit, which is composed of five platinum (Pt)-coated electrode plates, separated by semi-permeable membranes and the water is electrolyzed while passing through the gaps between the electrodes (Fig. 1C). Platinum-coated titanium electrodes are certified for at least 1,400 hours use without a marked deterioration with respect to the efficacy of water electrolysis, suggesting that the loss of a small amount of Pt nanoparticles from the surface of the electrode will not significantly affect the electrolysis efficacy of the device used here. Electrolyzed tap water near the cathode typically exhibits a high pH, low dissolved oxygen, high negative redox potential and a high concentration of dissolved hydrogen (0.4–0.9 ppm) (Table 1, [48]). Water produced in this manner, with the above characteristics, is designated as ERW. The test apparatus is designed to produce five types of water; four types of ERW (Levels 1–4) electrolyzed with a constant electric current for each level (0.8 to 4.2 A) at a maximum of 50 volts and one type of filtered water without electrolysis (Table 1). ERW is produced near the cathode, as indicated by the thick right-facing arrows in Fig. 1c, and positively charged radioactive Cs ions will be attracted to the cathode side during electrolysis, resulting in an increased concentration of Cs$^+$ ions in ERW, dependent upon the current intensity. Conversely, negatively charged I ions will be attracted to the anode side, resulting in a decreased concentration of I ions in ERW. The electrolysis currents were increased in the order of levels 1 to 4, where Level 4 represents the strongest current, reflecting the highest dissolved hydrogen (DH) and the lowest oxidation-reduction potential (ORP) (Table 1). When the radioactivity of ERW at level 4 is measured as being lower than the background level, then one can conclude that the radioactivity of ERW at levels 1 to 3 is lower than the background level. ERW at levels 1 to 3 is usually used for drinking and at level 4 is used for cooking. We have included Table 1 to aid the readers understanding of the four types of ERW.

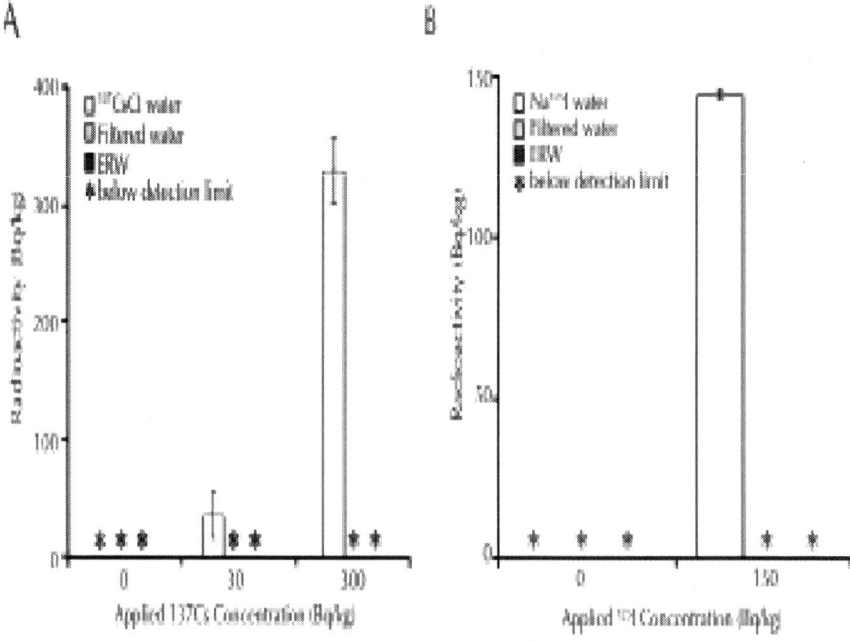

Figure 1: Schematic of the flow-type electrolysis apparatus. The test apparatus is composed of two units, a micro-carbon CM cartridge (B) and an electrolysis unit (C). The overall water flow and equipment set up is shown in (A).

Sample water is connected to an adjustable speed pump to maintain a flow rate of 1.8–2.0 l/min and expelled to the inlet of the electrolysis unit (A). Tap water passes through the nonwoven-fabric filter, the mixed layers of activated charcoal powders and cationic ion-exchange material to make filtered water (B). Filtered water then flows into the electrolysis unit composed of platinum-coated 5 electrode plates separated by semi-permeable membranes (C). Filtered water will be electrolyzed at levels 1, 2, 3 and 4 at a maximum of 50 volts while passing through the gaps between the electrodes.

Preparation of Non-Radioactive Sample Water (CsCl, Ki)

Tap water was used as a control. CsCl solutions of 20 liters each with concentrations of 20 and 2,000 ppb were prepared using tap water. Likewise, KI solutions with concentrations of 100 and 4,000 ppb were prepared. These solutions are designated as sample waters. The test system was arranged by placing an adjustable speed pump between the sample waters and the test apparatus to mimic tap water pressure, connected to the inlet of the test apparatus, as shown in Fig. 1A. The water flow rate was set to 1.8–2.0 L/min by adjusting the pump speed throughout the entire experiment. In the experiment, 1–2 liters of tap water was used to wash and equilibrate the system each time the sample concentrations were changed. Fifteen milliliters of filtered, ERW and relevant control waters were collected for ICP-MS analysis. The removal efficiency was calculated according to a previously described equation [38], shown in Tables 2 and 3.

Table 2: Removal efficiencies (%) for Cs ion and ^{137}Cs

Measured (Loaded) amounts		Removal efficiency (%)
as Cs ion (ppb)	as 137Cs (Bq/kg)	
1976.47 (2000)	0	58.2
20.55 (20)	0	87.4
*3.1600	16212.0 (15000)	96.9
*0.6360	3262.0 (3000)	96.9
*0.0642	329.0 (300)	99.2*

*0.0067	34.9 (30)	92.5*

Removal efficiency (%) = (12[A]/[B])6100 according to [38]. [A], [B]: concentrations of Cs and 137Cs after and before filtration. Each solution was filtered only, without electrolysis. *: [A] values were below the detection limit. *: [A] values used to calculate removal efficiency were below the detection limit. #: equivalent ppb values calculated from the radioactivities loaded. Values within parentheses were prepared and loaded amounts or radioactivities of cesium.

Table 3: Removal efficiencies (%) for I and ^{125}I ions

Measured (Loaded) amounts		Removal efficiency (%)
as I ion (ppb)	as 125I (Bq/kg)	
3891.0 (4000)	0	84.6
130.0 (100)	0	91.7
*0.0000197	14993.0 (15000)	99.4
*0.00000351	1788.0 (1500)	99.3
*0.000000196	146.3 (150)	99.5*

Removal efficiency (%) = (12[A]/[B])6100 according to [38]. [A], [B]: concentrations of I and 125I solutions after and before filtration. Each solution was filtered only, without electrolysis. *: [A] values used to calculate removal efficiency were below the detection limit. #: equivalent ppb values calculated from the radioactivities loaded. Values within parentheses were prepared and loaded amounts or radioactivities of iodine.

ICP-MS Analysis of Cs and I Elements in ERWs

Sample waters were passed through the apparatus, and collected filtered waters were quantitated using ICP-MS (Agilent 7500c, Agilent Technologies Co. Ltd., and Santa Clara, CA, USA) in the Radioisotope Center at Kyushu University.

Preparation of Radioactive Sample Water (^{137}CsCl and Na^{125}I)

Stock solution of ^{137}CsCl was diluted with 20 liters of tap water to prepare concentrations of 15,000, 3,000, 300, and 30 Bq/Kg. Likewise, Na^{125}I stock solution was diluted with 20 liters of tap water to prepare concentrations of 15,000, 1,500, and 150 Bq/Kg. All other experimental conditions, such as water flow rate, system equilibration, the electrolysis conditions of the apparatus were carried out as closely as possible to those used for the non-radioisotope experiments, except that 10 ml of each of the sample waters were collected for radioactivity counting.

Radioactivity Counting of ^{137}Cs and ^{125}I in Sample Waters

Radioactive sample waters were passed through the apparatus, and collected waters were quantitated using a gamma counter (AccuFLEX ARC-7001, Hitachi Aloka Medical, Ltd., Tokyo, Japan) in the Center of Advanced Instrumental Analysis at Kyushu University. To evaluate the effect of the electrolysis step on radionuclide removal, filtered waters were electrolyzed by a constant current (4.2 A) at level 4 and radioactivities of ERW were quantitated as above.

Statistical Analysis

All experiments were performed in triplicate. Data are expressed as means ± SD for each experiment.

RESULTS

Analysis of Cs and I Elements in the Filtered Water

Prior to radioisotope experiments, CsCl and KI solutions were prepared as described in the Materials and Methods section and their removability was tested. The background Cs concentration in tap water was similar to that for the filtered water (Fig. 2A, column 0 ppb). When 20 and 2,000 ppb CsCl solutions were used, the measured values of the filtered water indicate that the test apparatus had a higher removability (87.4%) for the 20 ppb CsCl solution than for the 2,000 ppb CsCl solution (58.2%) (Fig. 2A, Table 2). Similar experiments using KI solutions were carried out and the results are shown in Fig. 2B. The background I concentrations in tap water and that for the filtered water were similar (Fig. 2B, column 0 ppb). Removal efficiency after filtration for 100 ppb and 4,000 ppb KI solutions were 91.7% and 84.6%, respectively (Table 3). These results demonstrate that the micro-carbon CM cartridge is capable of removing Cs and I ions at all concentration ranges tested (Fig. 2).

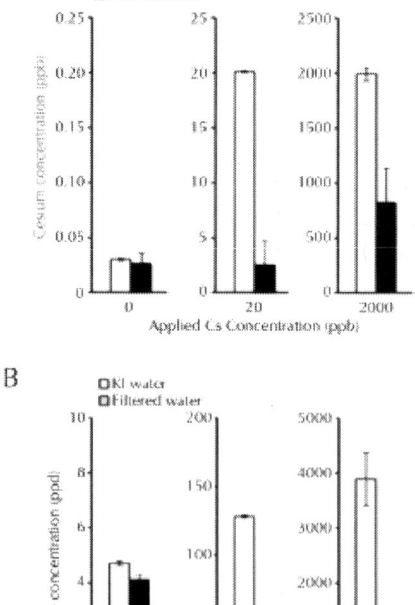

Figure 2: Measurement of Cs and I elements in filtered waters. CsCl solutions at concentrations of 0, 20 and 2,000 ppb were passed through the test ap-

paratus. Collected filtered waters were used to measure Cs concentration by ICP-MS (A). KI solutions at concentrations 0, 100 and 4,000 ppb were passed through the test apparatus. Collected filtered waters as in (A) were used to measure I concentration by ICP-MS (B). White bar: Tap water, gray bar: Filtered water. Experiments were carried out in triplicate.

Removal Efficiency of ^{137}CsCl and Na^{125}I in the Filtered Water

Because the test apparatus removed Cs and I ions efficiently, assays were extended to examine the removability of ^{137}CsCl and Na^{125}I. The natural background counts in tap water and filtered water were below the detection limit of the gamma counter (Fig. 3A and B, column 0). Tap water containing 30 (0.0067 ppb as Cs ions), 300 (0.0642 ppb as Cs ions), 3,000 (0.636 ppb as Cs ions) and 15,000 (3.16 ppb as Cs ions) Bq/kg of ^{137}CsCl as controls showed the expected radioactive counts (Fig. 3A and 3B, white bar at each concentration) with a high correlation coefficient (Fig. 3C, R^2 = 0.999). Control waters were then passed through the micro-carbon CM cartridge and the filtrate radioactivities were measured (Fig. 3A and B). It was found that the radioactivities of the filtered water for ^{137}CsCl were reduced significantly (Fig. 3A and 3B) and removal efficiency was 96.9%, even after loading 15,000 Bq/kg of ^{137}CsCl (Table 2).

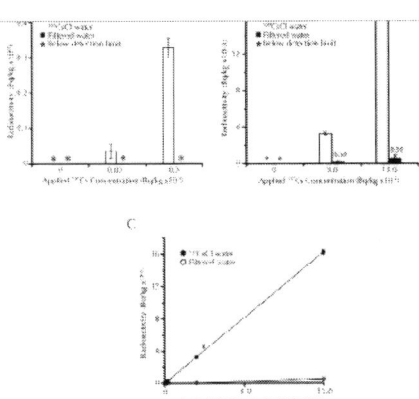

Figure 3: Measurement of ^{137}Cs in sample waters. ^{137}CsCl solutions at concentrations of 0, 0.03, 0.3, 3.0 and 15.0 KBq/kg were passed through the

test apparatus. Collected filtered waters were used to measure ^{137}Cs counts by an AccuFLEX ARC-7001 gamma counter (A and B). White bar: ^{137}CsCl solutions before filtration, gray bar: ^{137}CsCl solutions after filtration. Radioactivities before and after filtration were evaluated by linear-regression analysis (C). •: ^{137}CsCl solutions before filtration, o: ^{137}CsCl solutions after filtration. Experiments were carried out in triplicate.

To evaluate Na^{125}I removability, we prepared Na^{125}I containing sample waters as described above. The natural background count in tap water and filtered water exhibited values below the detection limit (Fig. 4A and B, column 0). Tap water containing 150 (0.000196 ppt as I ions), 1,500 (0.00351 ppt as I ions) and 15,000 (0.0197 ppt as I ions) Bq/kg of Na^{125}I as controls showed expected radioactive counts (Fig. 4A and 4B, white bar at each concentration) with a high correlation coefficient (Fig. 4C, R^2 = 0.999). Radioactive control tap waters were passed through the micro-carbon CM cartridge, reducing the filtrate radioactivities significantly (Fig. 4A and B), with a removal efficiency of over 99% (Table 3). Thus, the micro-carbon CM cartridge was demonstrated to efficiently remove radioactivities up to 15,000 Bq/kg of Na^{125}I.

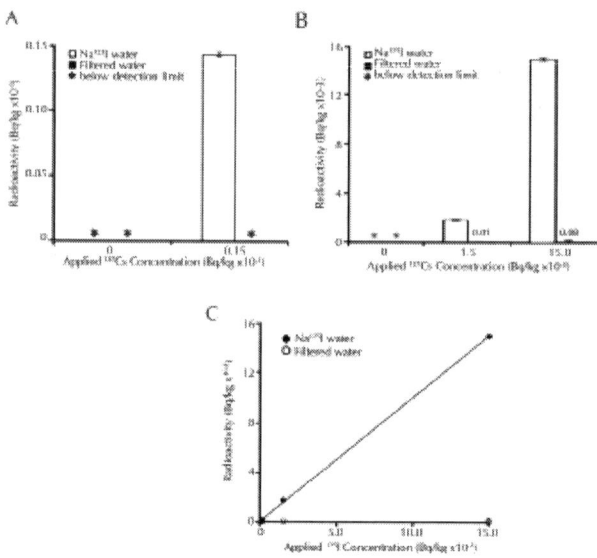

Figure 4: Measurement of ^{125}I elements in sample waters. Na^{125}I solutions at concentrations of 0, 0.15, 1.5 and 15.0 KBq/kg were passed through the

natural and impregnated zeolite minerals J Hazard Mater. 172: 416–422. doi: 10.1016/j.jhazmat.2009.07.033

16. Shakir K, Sohsah M, Soliman M (2007) Removal of cesium from aqueous solutions and radioactive waste simulants by coprecipitate flotation. Sep Purif Technol 54: 373–381. doi: 10.1016/j.seppur.2006.10.006

17. Osmanlioglu AE (2007) Natural diatomite process for removal of radioactivity from liquid waste. Appl Radiat Isot 65: 17–20. doi: 10.1016/j.apradiso.2006.08.012

18. Hu B, Fugetsu B, Yu H, Abe Y (2012) Prussian blue caged in spongiform adsorbents using diatomite and carbon nanotubes for elimination of cesium. J Hazard Mater 217–218: 85–91. doi: 10.1016/j.jhazmat.2012.02.071

19. Vipin AK, Hu B, Fugetsu B (2013) Prussian blue caged in alginate/calcium beads as adsorbents for removal of cesium ions from contaminated water. J Hazard Mater 258–259: 93–101. doi: 10.1016/j.jhazmat.2013.04.024

20. Thammawong C, Opaprakasit P, Tangboriboonrat P, Sreearunothai P (2013) Prussian blue-coated magnetic nanoparticles for removal of cesium from contaminated environment. J Nanopart Res 15: 1689. doi: 10.1007/s11051-013-1689-z

21. Dahiya S, Tripathi RM, Hegde AG (2008) Biosorption of heavy metals and radionuclide from aqueous solutions by pre-treated arca shell biomass. J Hazard Mater 150: 376–386. doi: 10.1016/j.jhazmat.2007.04.134

22. Pangeni B, Paudyal H, Inoue K, Ohto K, Kawakita H, et al. (2014) Preparation of natural cation exchanger from persimmon waste and its application for the removal of cesium from water. Chem Eng J 242: 109–116. doi: 10.1016/j.cej.2013.12.042

23. Ding D, Lei Z, Yang Y, Feng C, Zhang Z (2014) Selective removal of cesium from aqueous solutions with nickel (II)hexacyanoferrate (III) functionalized agricultural residue–walnut shell J Hazard Mater. 270: 187–195. doi: 10.1016/j.jhazmat.2014.01.056

24. Awual MR, Suzuki S, Taguchi T, Shiwaku H, Okamoto Y, et al. (2014) Radioactive cesium removal from nuclear wastewater by novel inorganic and conjugate adsorbents. Chem Eng J 242: 127–135. doi: 10.1016/j.cej.2013.12.072

6. Thakur P, Ballard S, Nelson R (2013) an overview of Fukushima radionuclides measured in the northern hemisphere. Sci Total Environ 458–460: 577–613. doi: 10.1016/j.scitotenv.2013.03.105
7. Murakami M, Ohte N, Suzuki T, Ishii N, Igarashi1 Y, et al (2014) Biological proliferation of cesium-137 through the detrital food chain in a forest ecosystem in Japan. Sci Rep 4: 3599. doi: 10.1038/srep03599
8. Nakanishi T, Matsunaga T, Koarashi J, Atarashi-Andoh M (2014) ^{137}Cs vertical migration in a deciduous forest soil following the Fukushima Dai-ichi Nuclear Power Plant accident. J Environ Radioact 128: 9–14. doi: 10.1016/j.jenvrad.2013.10.019
9. Momoshima N, Sugihara S, Ichikawa R, Yokoyama H (2012) Atmospheric radionuclides transported to Fukuoka, Japan remote from the Fukushima Dai-ichi nuclear power complex following the nuclear accident. J Environ Radioact 111: 28–32. doi: 10.1016/j.jenvrad.2011.09.001
10. Parajuli D, Tanaka H, Hakuta Y, Minami K, Fukuda S, et al. (2013) Dealing with the Aftermath of Fukushima Daiichi Nuclear Accident: Decontamination of Radioactive Cesium Enriched Ash. Environ Sci Technol 47: 3800–3806. doi: 10.1021/es303467n
11. Liu X, Chen G-R, Lee D-J, Kawamoto T, Tanaka H, et al. (2014) Adsorption removal of cesium from drinking waters: A mini review on use of biosorbents and other adsorbents. Bioresour Technol Available: http://dx.doi.org/10.1016/j.biortech. Last accessed 2014.01.012.
12. Rana D, Matsuura T, Kassim MA, Ismail AF (2013) radioactive decontamination of water by membrane processes – A review. Desalination 321: 77–92. doi: 10.1016/j.desal.2012.11.007
13. Song K-C, Lee HK, Moon H, Lee KJ (1997) Simultaneous removal of the radiotoxic nuclides Cs^{137} and I^{129} from aqueous solution. Sep Purif Technol 12: 215–227. doi: 10.1016/s1383-5866(97)00045-2
14. El-Kamash AM (2008) Evaluation of zeolite A for the sorptive removal of Cs$^+$ and Sr^{2+}ions from aqueous solutions using batch and fixed bed column operations. J Hazard Mater 151: 432–445. doi: 10.1016/j.jhazmat.2007.06.009
15. Borai EH, Harjula R, Malinen L, Paajanen A (2009) Efficient removal of cesium from low-level radioactive liquid waste using

ACKNOWLEDGMENTS

The authors thank Nihon Trim Co. Ltd. for providing the Trim Ion NEO apparatus and an adjustable flow rate pump. The authors are also grateful to Ms. Yuri Fujimoto and Chika Kubota for their technical assistance.

AUTHOR CONTRIBUTIONS

Conceived and designed the experiments: TH NN KT SS. Performed the experiments: TH NN SS. Analyzed the data: TH NN KT SS. Contributed reagents/materials/analysis tools: TH NN SS. Wrote the paper: TH NN SS.

REFERENCES

1. Hamada N, Ogino H (2012) Food safety regulations: what we learned from the Fukushima nuclear accident. J Environ Radioact 111: 83–99. doi: 10.1016/j.jenvrad.2011.08.008
2. Hamada N, Ogino H, Fujimichi Y (2012) Safety regulations of food and water implemented in the first year following the Fukushima nuclear accident. J Radiat Res 53: 641–671. doi: 10.1093/jrr/rrs032
3. Tagami K, Uchida S (2011) Can we remove iodine-131 from tap water in Japan by boiling? – Experimental testing in response to the Fukushima Daiichi Nuclear Power Plant accident. Chemosphere 84: 1282–1284. doi: 10.1016/j.chemosphere.2011.05.050
4. Amano H, Akiyama M, Chunlei B, Kawamura T, Kishimoto T, et al. (2012) Radiation measurements in the Chiba Metropolitan Area and radiological aspects of fallout from the Fukushima Dai-ichi Nuclear Power Plants accident. J Environ Radioact 111: 42–52. doi: 10.1016/j.jenvrad.2011.10.019
5. Koizumi A, Niisoe T, Harada KH, Fujii Y, Adachi A, et al. (2013) ^{137}Cs trapped by biomass within 20 km of the Fukushima Daiichi Nuclear Power Plant. Environ Sci Technol 47: 9612–9618. doi: 10.1021/es401422g

ERW is regarded as beneficial to health because of its ROS scavenging ability [39]. ERW produced from tap water by this apparatus could contain as much as 0.587 ppm of dissolved hydrogen (Table 1, [48]). This hydrogen concentration in ERW is relatively high for a flow-type electrolysis apparatus when compared with the concentration of 1.6 ppm hydrogen in 100% hydrogen-saturated water [53]. Such dissolved molecular hydrogen has been shown to exert radioprotective effects in both *in vitro* and *in vivo* studies [49]–[53]. Molecular hydrogen in ERW prepared from tap water suppressed neuroinflammation in mice [48], and extended the life span of *C. elegans* [54]. Additionally, molecular hydrogen was demonstrated to act as a neuroprotective agent and ROS scavenger [67]. Moreover, ERW produced from an electrolysis unit incorporating Pt-electrodes has been shown to contain 0.1–0.25 ppb Pt nanoparticles [39],[54], [68]. Pt nanoparticles exhibit protective effects that are attributed to their suppressing ROS production caused by UV-light-induced epidermal inflammation [69]. Synthetic Pt nanoparticles have been shown to scavenge ROS in cultured HeLa cells [70], to induce expression of antioxidant enzyme genes in rat skeletal muscle L6 cells [71], and to act as an SOD/catalase mimetic agent in human lymphoma cells [72]. Model ERW prepared from NaCl, KCl or NaOH solutions has been shown to exert beneficial effects such as anti-diabetic, anti-cancer, and life-span extension of nematodes because of its ROS scavenging ability in numerous *in vitro* and *in vivo* studies [73]–[78]. Therefore, molecular hydrogen and Pt nanoparticles dissolved in ERW could synergistically contribute to protect gastrointestinal damage caused by ingested radioactive foodstuffs. Furthermore, to maximize protective efficacy against radiation-induced gastrointestinal damage, the consumption of various supplemental foods such as naringin [42], probiotics [57], [66], Kefir [79], melatonin [80] and curcumin [81] are reported to be beneficial.

In conclusion, we demonstrated that radio-cesium and -iodine are efficiently removed by an apparatus containing a micro-carbon CM cartridge filter, prior to ingestion. We also suggest that the ERW produced by the test apparatus will provide maximum protection against accidentally and/or unconsciously ingested radionuclides because it contains dissolved hydrogen and Pt nanoparticles. Therefore, the test apparatus is considered to be a potential alternative tool to minimize radiation hazards caused by contaminated foodstuffs.

Only experiments using low levels of radionuclides will answer the question of whether such interactions between the constituents and added radionuclides may affect removability by this apparatus. Another reason to use lower levels of radionuclides is that even a small amount of ^{137}Cs dissolved in water is difficult to remove [11], [20] and accumulates in the body, causing prolonged exposure. Moreover, the fact is that low levels of radioactive Cs species currently contaminate drinking water in many cities around FDNPP [36]. This may be partly attributed to the limited removability of solubilized cesium by the conventional coagulation-sedimentation process [11],[29]. It is therefore extremely important, for the residents of affected regions, to find a way to remove even small amounts of nuclear contaminants from drinking water.

Another concern related to radiocesium is its longer half-life and a characteristic of ready transfer to the human diet through plants [62]. Precipitated Cs$^+$ binds to clay minerals rather tightly [27], and depth distribution studies reveal that approximately 80% of total radiocesium is retained in the upper 2.0 cm of tested soil samples [63]. Another study estimated that ^{137}Cs could reach a depth of only 18 cm after 300 yr [37]. These characteristics of surface area retention of radiocesium in addition to its long physical half-life (^{134}Cs, $T_{1/2}$ = 2.06 yr; ^{137}Cs, $T_{1/2}$ = 30.17 yr) could be a secondary contamination source for vegetation via roots. Uptake of radiocesium from the root is thought to occur via the potassium transport system and is distributed rapidly within the plants [62]. Indeed, many agricultural products are reported to be contaminated by radiocesium and are their marketing is restricted [64], [65]. Ingestion of radiocesium-contaminated foodstuffs will expose the gastrointestinal tract and be absorbed into tissues and organs in the body. Gastrointestinal, reproductive and hematopoietic systems are sensitive to ionizing radiation due to their high turnover rate [57], [66]. As an example, the degeneration of small intestinal mucosa cells is caused by free radicals produced from the interactions of radiation energy with intracellular water molecules [66]. Water radiolysis generates a variety of ROS that cause extensive oxidative damages to biologically critical macromolecules, leading to cell death [40], [42]–[45]. Therefore, providing a method to counter radiation hazards caused by accidentally ingested radioactive waters and foodstuffs will be a great contribution to human health.

2011 in Tamura, Fukushima Prefecture. It was also reported that a sum of ^{134}Cs and ^{137}Cs less than 32 Bq/kg was sporadically detected in tap water during 22 days of monitoring after the accident [29]. The water purification plants take precautions not to distribute contaminated water through constant monitoring to meet the latest upper limit value, set by the government as 10 Bq/kg for drinking water, effective from April 1, 2012 [56]. In reality, the detection of greater than 10 Bq/kg radioactivities in tap water in general public is most likely to be the result of accidental and sporadic contamination events. In any case, the test apparatus was demonstrated to decontaminate radiocesium levels to below the detection limit, even when tap water was contaminated by up to 300 Bq/kg radiocesium. When loading 300 Bq/kg of ^{137}Cs to the cartridge, the removability obtained was a conditional value of 99.2% and leaving the remaining radioactivity to be below the upper limit value of 10 Bq/kg set by the Government. This indicates that the filtered water right before entering into the electrolysis unit still contains a trace amount of ^{137}Cs and the following electrolysis step may produce ^{137}Cs enriched ERW. The test apparatus is a powerful electrolysis device yet finely tuned to produce various levels of dissolved hydrogen electric current dependently (Table 1). Under these conditions, we could not definitely exclude a slight possibility that the electrolysis step contributes to ^{137}Cs enrichment in ERW. Thus, the only way to clarify such uncertainty was to conduct the experiments as shown in Fig. 5. Moreover, we judged that it is not sufficient enough by just showing the removability of the cartridge filter unit alone and extrapolating the results for evaluating the entire flow-type system. To this end, we decided to measure the radioactivity in ERW, which allows evaluating cartridge unit and electrolysis unit simultaneously. Therefore, the evaluation of the filtering unit in combination with the electrolysis unit as the complete flow-type system was necessary. Our concern for negatively charged I ion was less intense compared to Cs^+ ion due to higher removability by the cartridge unit and attracted to the anode side. Nevertheless, we confirmed the removability of the flow-type system experimentally to provide the data set with Cs^+ data.

It is commonly regarded that tap water prepared from lakes and rivers contains varying amounts of organic and inorganic materials. In the present study, we considered these to have a significant impact on removability by the test system because such materials are most likely to compete with the very small amounts of ^{137}Cs and ^{125}I ions present.

hollow fiber membranes do not contribute to the elimination of iodate (IO_3^-) ions because their radius is 0.326 nm, even when their radius is increased several fold in water [61]. Additionally, the ineffectiveness of removing ^{131}I by boiling tap water has been reported [3].

Cesium is an alkaline earth metal that exists as a monovalent cation form (Cs^+) in water and in soils [27]. We found that Cs^+ could be efficiently removed by the micro-carbon CM cartridge tested here. The mechanism for the removal of Cs^+ remains to be investigated. The Cs^+ removal efficiency by the apparatus was 87.4% at 20 ppb, which is comparable to that of several pot-type water purifiers that have efficiencies of around 90% for tap water containing 40–50 μg/l (ppb) cesium chloride [38]. A removal efficiency of 58.2% for Cs^+ appears to be low at the highest concentration (1976.5 ppb) loading. This lower removal efficiency could be explained by the amount of Cs^+ getting close to system over loading because this amount is 625.3 times more Cs^+ ion loading than the 3.16 ppb Cs^+ ion calculated from the highest radioactivity (16,212 Bq/kg) loading where 96.9% removability was attained (Table 2). Therefore, the apparatus could remove ^{137}Cs with above 96% efficiency for less than a 3.16 ppb CsCl loading and the removal efficiency is higher than that reported for two commercialized pot-type water purifiers, composed of activated charcoal and an ion exchanger, or activated charcoal, ceramics and a hollow fiber membrane, with 84.2–91.5% efficiencies for rain water samples[27]. Another set of experiments using commercialized four pot-type purifiers made of materials similar to those above assessed iodine and cesium removability, with efficiencies of approximately 85% and 75–90%, respectively [38]. Others also tested Cs removability using a spongiform adsorbent made of Prussian blue caged within the diatomite cavities and carbon nanotubes, by contacting for 10 hours with low levels of ^{137}Cs, yielding a 99.93% removal efficiency [18]. The present test apparatus showed a removal efficiency of over 96% for Cs and I, which is competitive with or better than previously reported removal efficiencies ranging from 75% to 99.93%. It is emphasized here that the advantages of the test apparatus are that it has long been used for domestic use, is easy to operate, provides a sufficient amount of purified water instantaneously (max. 5 l/min.) and offers an established system for proper disposal and/or recycling of used cartridges. Following the FDNPP accident, tap water contamination monitoring revealed that the maximum of the sum of ^{134}Cs and ^{137}Cs was 180.5 Bq/kg on March by reacting with chlorine [29], and as a result, almost all iodine is converted to the iodate ion (IO_3^-) in tap water due to the oxidation by chlorine [59]. It is reported that $^{131}I^-$ removal is increased by water containing 0.1–0.5 mg/L chlorine, with lower concentrations of powdered activated charcoal [29]. However, granular and powdered activated carbons were reported to remove ^{131}I at about 30–40% efficiency. Additionally, it has been reported that $^{125}I^-$ and $^{125}I_3^-$ were prepared from ^{125}I and used to test the removability of these species by a granular type charcoal, which resulted in a small amount of adsorption [60]. These results may partly explain the inefficient removability by activated charcoal reported by others, through selective adsorption of iodate and iodine [38], [60], [61]. Activated carbon was shown to remove iodide (I^-) more efficiently than iodate (IO_3^-) [27]. Therefore, it appears that combinations between the types of activated carbon/charcoal and iodine species affect overall removability. In the present experiments, we used tap water distributed by the Waterworks Bureau of the City of Fukuoka, expected to contain at least 0.1 mg/L chlorine. Thus, ^{125}I is mostly, if not completely, converted to iodate ions (IO_3^-) by chlorine in the tap water. In the present results, KI and ^{125}I were efficiently removed from tap waters by the micro-carbon CM cartridge, suggesting that iodide and iodate ions were removed. The micro-carbon CM cartridge is composed of a nonwoven-fabric filter and activated carbons consisting of a coconut shell activated carbon powder, a coconut shell activated carbon conjugated with a silver compound for antimicrobial effect, and an amorphous titanosilicate-based inorganic compound (BASF Co, Germany) molded with a fibrous binder for shaping. This cartridge was used in the present test apparatus to remove particulate matters, microorganisms, and for qualified removability of 13 designated impurities, tested according to the standard method set by JIS S 3201, 2004 (Domestic Water Purifier Quality Test) [55]. It is worth mentioning that the test apparatus effectively removed I and ^{125}I (applicable to Cs^+ and ^{137}Cs), even though water was supplied to the apparatus through a pump simulating tap water outlet pressure to attain 1.8–2.0 L/min flow rate, which markedly reduced the contact time of water with the activated carbon surfaces and ion-exchangers compared with those in pot-type water purifiers. It has been reported that the above-mentioned molded activated carbons can replace the hollow fiber membrane filter that is commonly used in other water purifiers to eliminate materials larger than 0.1 μm in size [55]. Incidentally,

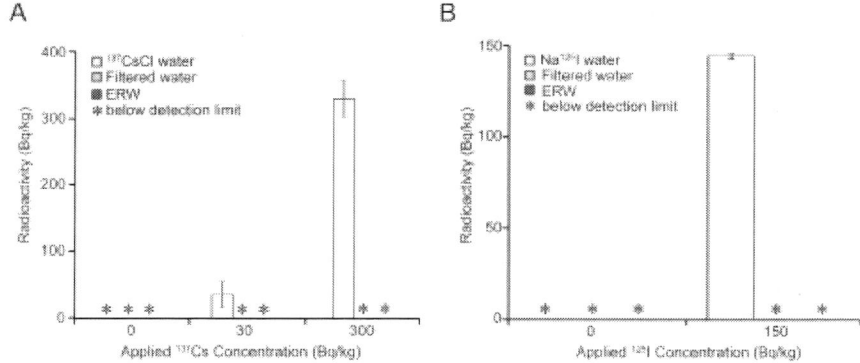

Figure 5: Effects of electrolysis on filtered radioactive sample waters. ^{137}CsCl solutions of 30 and 300 Bq/kg were passed through the test apparatus and filtered waters were collected for measurement. Then, filtered water was passed through the electrolysis unit at the highest electrolysis level of 4 and ERW was collected for measurement. Collected waters were used to measure ^{137}Cs counts by an AccuFLEX ARC-7001 gamma counter (A and B). Using the same protocol, filtered water and ERW were collected for 150 Bq/kg of Na^{125}I solution. White bar: ^{137}CsCl or Na^{125}I solutions, gray bar: Filtered ^{137}CsCl or Na^{125}I solutions; black bar: ERWs of filtered ^{137}CsCl or Na^{125}I solutions. Experiments were carried out in triplicate.

DISCUSSION

The FDNPP accident liberated various radionuclides, including ^{131}I, ^{132}I, ^{134}Cs, and ^{137}Cs[57]. Amongst these radionuclides, ^{131}I can enter the body through inhalation and by ingesting contaminated foodstuffs including drinking water, which then rapidly concentrates in the thyroid gland, where -radiation exposure takes place. As its half-life is 8 days, radioactivity levels are expected to be reduced substantially over several months. Therefore, an obvious precaution is not to ingest ^{131}I-contaminated or doubtful foodstuffs including drinking water. Water supply law in Japan limits the lowest chlorine concentration in tap water outlet at 0.1 mg/L [58]. Dissolved ^{131}I is reported to form various species in tap water such as the radioactive iodide ion (^{131}I$^-$), hypoiodous acid (HO^{131}I), the iodate ion (^{131}IO$_3^-$), iodine molecules (^{131}IO$_2$) and organic ^{131}I. ^{131}I$^-$ reacts with chlorine and is transformed mainly into HOI at neutral pH. HOI is further transformed into IO$_3^-$

test apparatus. Collected filtered waters were used to measure ^{125}I counts by an AccuFLEX ARC-7001 gamma counter (A and B). White bar: Na^{125}I solutions before filtration, gray bar: Na^{125}I solutions after filtration. Radioactivities before and after filtration were evaluated by linear-regression analysis (C). •: Na^{125}I solutions before filtration, ○: Na^{125}I solutions after filtration. Experiments were carried out in triplicate.

Effect of Electrolysis on the Removal Efficiency of ^{137}Cs and ^{125}I

In parallel with the preceding experiments, we evaluated the effects of the electrolysis step in terms of efficiencies for ^{137}Cs and ^{125}I removal from the filtered radioactive water. Filtered water was electrolyzed at the highest current level of 4. In this experiment, we selected 300 Bq/kg of ^{137}Cs water, which loaded 30 times more radioactivity than the upper limit value of 10 Bq/kg for drinking water set by the government [56]. Under these conditions, the radioactivity in ERW remained below the detection limit (Fig. 5A). Similarly, we evaluated the removability of ^{125}I by the highest electrolysis level of 4. In this case, we selected 150 Bq/kg of ^{125}I, which is a loading of 1.5 times more radioactivity than the upper limit of 100 Bq/L of ^{131}I concentration for infants under 1 year of age set by the Ministry of Health, Labour and Welfare, 1947 [37]. The radioactive iodine level in ERW remained below the detection limit (Fig. 5B). Therefore, the results indicate that the cartridge substantially removed ^{137}Cs and ^{125}I from tap water prior to the electrolysis step, thereby assuring undetectable levels of radioactivity in ERW produced at the highest current level of 4, which has the highest attraction for ^{137}Cs$^+$, and thus the results hold true for ERWs produced with the current levels 1 to 3.

25. Valsala TP, Roy SC, Shah JG, Gabriela J, Raj K, et al. (2009) Removal of radioactive caesium from low level radioactive waste (LLW) streams using cobalt ferrocyanide impregnated organic anion exchanger. J Hazard Mater 166: 1148–1153. doi: 10.1016/j.jhazmat.2008.12.019
26. Sengupta P, Dudwadkar NL, Vishwanadh B, Pulhani V, Rao Rekha, et al. (2014) Uptake of hazardous radionuclides within layered chalcogenide for environmental protection. J Hazard Mater 266: 94–101. doi: 10.1016/j.jhazmat.2013.12.010
27. Higaki S, Hirota M (2012) Decontamination Efficiencies of Pot-Type Water Purifiers for ^{131}I, ^{134}Cs and ^{137}Cs in Rainwater Contaminated during Fukushima Daiichi Nuclear Disaster. PLoS ONE 7(5): e37184. doi: 10.1371/journal.pone.0037184
28. Goossens R, Delville A, Genot J, Halleux R, Masschelein WJ (1989) Removal of the typical isotopes of the Chernobyl fall-out by conventional water treatment. War Res 23: 693–697. doi: 10.1016/0043-1354(89)90201-7
29. Kosaka K, Asami M, Kobashigawa N, Ohkubo K, Terada H, et al. (2012) Removal of radioactive iodine and cesium in water purification processes after an explosion at a nuclear power plant due to the Great East Japan Earthquake. Water res 46: 4397–4404. doi: 10.1016/j.watres.2012.05.055
30. Shimura H, Itoh K, Sugiyama A, Ichijo S, Ichijo M, et al. (2012) Absorption of Radionuclides from the Fukushima Nuclear Accident by a Novel Algal Strain. PLoS ONE 7(9): e44200. doi: 10.1371/journal.pone.0044200
31. Fukuda S, Iwamoto K, Atsumi M, Yokoyama A, Nakayama T, et al. (2014) Global searches for microalgae and aquatic plants that can eliminate radioactive cesium, iodine and strontium from the radio-polluted aquatic environment: a bioremediation strategy. J Plant Res 127: 79–89. doi: 10.1007/s10265-013-0596-9
32. Szabo J, Minamyer S (2014) Decontamination of radiological agents from drinking water infrastructure: A literature review and summary. Environ Int Available:http://dx.doi.org/10.1016/j.envint.2014.01.020.
33. Ministry of Health, Labour and Welfare, Japan. Information on the Great East Japan Earthquake-Water supply. Available: www.

mhlw.go.jp/english/topics/2011eq/index.html. Accessed Jul. 18, 2013.
34. Ministry of Health, Labour and Welfare, Japan. Measurement results of radionuclides in foodstuffs (No. 522) Available: http://www.mhlw.go.jp/stf/houdou/2r98520 00002oo2l-att/2r9852000002oo6v.pdf. Accessed Jul. 18, 2013.
35. Mizuno T, Kubo H (2013) Overview of active cesium contamination of fresh water fish in Fukushima and Eastern Japan. Sci Rep 3: 1742. doi: 10.1038/srep01742
36. Nuclear Regulation Authority (NRA). Monitoring information of environmental radioactivity level: Readings of radioactivity level in drinking water by prefecture October–December, 2013, Accessed Mar. 13, 2014.
37. Ohta T, Mahara Y, Kubota T, Fukutani S, Fujiwara K, et al. (2012) Prediction of groundwater contamination with ^{137}Cs and ^{131}I from the Fukushima nuclear accident in the Kanto district. J Environ Radioact 111: 38–41. doi: 10.1016/j.jenvrad.2011.11.017
38. Sato I, Kudo H, Tsuda S (2011) Removal efficiency of water purifier and adsorbent for iodine, cesium, strontium, barium and zirconium in drinking water. J Toxicol Sci 36(6): 829–834. doi: 10.2131/jts.36.829
39. Shirahata S, Hamasaki T, Teruya K (2012) Advanced research on the health benefit of reduced water. Trends Food Sci Technol 23: 124–131. doi: 10.1016/j.tifs.2011.10.009
40. Ewing D, Jones SR (1987) Superoxide Removal and Radiation Protection in Bacteria. Arch Biochem Biophys 254(1): 53–62. doi: 10.1016/0003-9861(87)90080-4
41. Ward JF (1988) DNA damage produced by ionizing radiation in mammalian cells: identities, mechanisms of formation, and reparability. Prog Nucleic Acid Res Mol Biol 35: 95–125. doi: 10.1016/s0079-6603(08)60611-x
42. Jagetia GC, Reddy TK (2005) Modulation of radiation-induced alteration in the antioxidant status of mice by naringin. Life Sci 77: 780–794. doi: 10.1016/j.lfs.2005.01.015
43. Nunomura A, Honda K, Takeda A, Hirai K, Zhu X, et al. (2006) Oxidative damage to RNA in neurodegenerative diseases. J Biomed Biotechnol 2006: Article ID 82323: 1–6. doi: 10.1155/jbb/2006/82323

44. Tanaka M, Chock PB, Stadtman ER (2007) Oxidized messenger RNA induces translation errors. Proc Natl Acad Sci USA 104(1): 66–71. doi: 10.1073/pnas.0609737104
45. Radak Z, Zhao Z, Goto S, Koltai E (2011) Age-associated neurodegeneration and oxidative damage to lipids, proteins and DNA. Mol Aspects Med 32: 305–315. doi: 10.1016/j.mam.2011.10.010
46. Cerutti PA (1985) Prooxidant states and tumor promotion. Science 227: 375–381. doi: 10.1126/science.2981433
47. Gobbel GT, Bellinzona M, Vogt AR, Gupta N, Fike John R, et al. (1998) Response of postmitotic neurons to X-irradiation: implications for the role of DNA damage in neuronal apoptosis. J Neurosci 18(1): 147–155.
48. Spulber S, Edoff K, Hong L, Morisawa S, Shirahata S, et al. (2012) Molecular hydrogen reduces LPS-induced neuroinflammation and promotes recovery from sickness behaviour in mice. PLoS ONE 7(7): e42078. doi: 10.1371/journal.pone.0042078
49. Qian L, Cao F, Cui J, Wang Y, Huang Y, et al. (2010) The potential cardioprotective effects of hydrogen in irradiated mice. J Radiat Res 51: 741–747. doi: 10.1269/jrr.10093
50. Qian L, Cao F, Cui J, Huang Y, Zhou X, et al. (2010) Radioprotective effect of hydrogen in cultured cells and mice. Free Radic Res 44(3): 275–282. doi: 10.3109/10715760903468758
51. Terasaki Y, Ohsawa I, Terasaki M, Takahashi M, Kunugi S, et al. (2011) Hydrogen therapy attenuates irradiation-induced lung damage by reducing oxidative stress. Am J Physiol Lung Cell Mol Physiol 301: L415–L426. doi: 10.1152/ajplung.00008.2011
52. Chuai Y, Gao F, Li B, Zhao L, Qian L, et al. (2012) Hydrogen-rich saline attenuates radiation-induced male germ cell loss in mice through reducing hydroxyl radicals. Biochem J 442: 49–56. doi: 10.1042/bj20111786
53. Ohno K, Ito M, Ichihara M, Ito M (2012) Molecular hydrogen as an emerging therapeutic medical gas for neurodegenerative and other diseases. Oxid Med Cell Longev 2012: Article ID 353152.
54. Yan H, Tian H, Kinjo T, Hamasaki T, Tomimatsu K, et al. (2010) Extension of the lifespan of *Caenorhabditis elegans* by the use of electrolyzed reduced water. Biosci Biotechnol Biochem 74(10): 2011–2015. doi: 10.1271/bbb.100250

55. Yoshinobu H, Arita S, Kawasaki S (2012) Molded activated charcoal and water purifier involving SAME. Patent application number: US20120132578.
56. Ministry of Health, Labour and Welfare, Japan. Available:http://www.mhlw.go.jp/shinsai_jo uhou/dl/leaflet_120329.pdf. Accessed Jul 18, 2013.
57. Christodouleas JP, Forrest RD, Ainsley CG, Tochner Z, Hahn SM, et al. (2011) Short-term and long-term health risks of nuclear-power-plant accidents. N Engl J Med 364: 2334–41. doi: 10.1056/nejmra1103676
58. Ministry of Health, Labour and Welfare, Japan. Available:http://www.mhlw.go.jp/shingi/2002/10/s1007-5c.html. Accessed Jul 18, 2013.
59. Kametani K, Matsumura T, Naito M (1992) Separation of iodide and iodate by anion exchange resin and determination of their ions in surface water (In Japanese). Bunseki Kagaku 41: 337–341. doi: 10.2116/bunsekikagaku.41.7_337
60. Watari K, Imai K, Ohmomo Y, Muramatsu Y, Nishimura Y, et al. (1988) Simultaneous adsorption of Cs-137 and I-131 from water and milk on "metal ferrocyanide-anion exchange resin". J Nucl Sci Technol 25(5): 495–499. doi: 10.1080/18811248.1988.9733618
61. Kamei D, Kuno T, Sato S, Nitta K, Akiba T (2012) Impact of the Fukushima Daiichi Nuclear Power Plant accident on hemodialysis facilities: An evaluation of radioactive contaminants in water used for hemodialysis. Ther Apher Dial1 6(1): 87–90. doi: 10.1111/j.1744-9987.2011.01029.x
62. Zhu Y-G, Smolders E (2000) Plant uptake of radiocaesium: a review of mechanisms, regulation and application. J Exp Bot 51(351): 1635–1645. doi: 10.1093/jexbot/51.351.1635
63. Kato H, Onda Y, Teramage M (2012) Depth distribution of ^{137}Cs, ^{134}Cs, and ^{131}I in soil profile after Fukushima Dai-ichi Nuclear Power Plant accident. J Environ Radioact 111: 59–64. doi: 10.1016/j.jenvrad.2011.10.003
64. Ministry of Health, Labour and Welfare (2013–595): Available:http://www.mhlw.go.jp/stf/houdou/2r98520 00002wvi2.html.Accessed Jul. 18, 2013.

65. Ministry of Agriculture, Forestry and Fisheries: Available:http://www.maff.go.jp/j/kanbo/joho/saigai/s_chosa/hinmoku_kekka.html.Accessed Jul. 18, 2013.
66. Spyropoulos BG, Misiakos EP, Fotiadis C, Stoidis CN (2011) Antioxidant properties of probiotics and their protective effects in the pathogenesis of radiation-induced enteritis and colitis. Dig Dis Sci 56: 285–294. doi: 10.1007/s10620-010-1307-1
67. Ohsawa I, Ishikawa M, Takahashi K, Watanabe M, Nishimaki K, et al. (2007) Hydrogen acts as a therapeutic antioxidant by selectively reducing cytotoxic oxygen radicals. Nat Med 13(6): 688–694. doi: 10.1038/nm1577
68. Yan H, Kinjo T, Tian H, Hamasaki T, Teruya K, et al. (2011) Mechanism of the lifespan extension of *Caenorhabditis elegans* by electrolyzed reduced water Participation of Pt nanoparticles. Biosci Biotechnol Biochem 75(7): 1295–1299. doi: 10.1271/bbb.110072
69. Yoshihisa Y, Honda A, Zhao Q-L, Makino T, Abe R, et al. (2010) Protective effects of platinum nanoparticles against UV-light-induced epidermal inflammation. Exp Dermatol 19: 1000–1006. doi: 10.1111/j.1600-0625.2010.01128.x
70. Hamasaki T, Kashiwagi T, Imada T, Nakamichi N, Aramaki S, et al. (2008) Kinetic analysis of superoxide anion radical-scavenging and hydroxyl radical-scavenging activities of platinum nanoparticles. Langmuir 24: 7354–7364. doi: 10.1021/la704046f
71. Nakanishi H, Hamasaki T, Kinjo T, Yan Hanxu, Nakamichi N, et al. (2013) Low concentration platinum nanoparticles effectively scavenge reactive oxygen species in rat skeletal L6 cells. Nano Biomed Eng 5(2): 76–85. doi: 10.5101/nbe.v5i2.p76-85
72. Yoshihisa Y, Zhao Q-L, Hassan MA, Wei Z-L, Furuichi M, et al. (2011) SOD/catalase mimetic platinum nanoparticles inhibit heat-induced apoptosis in human lymphoma U937 and HH cells. Free Radic Res 45(3): 326–335. doi: 10.3109/10715762.2010.532494
73. Li Y, Hamasaki T, Nakamichi N, Kashiwagi T, Komatsu T, et al. (2011) Suppressive effects of electrolyzed reduced water on alloxan-induced apoptosis and type 1 diabetes mellitus. Cytotechnology 63: 119–131. doi: 10.1007/s10616-010-9317-6

74. Kim M-J, Kim HK (2006) Anti-diabetic effects of electrolyzed reduced water in streptozotocin-induced and genetic diabetic mice. Life Sci 79: 2288–2292. doi: 10.1016/j.lfs.2006.07.027
75. Li Y, Nishimura T, Teruya K, Maki T, Komatsu T, et al. (2002) Protective mechanism of reduced water against alloxan-induced pancreatic -cell damage: Scavenging effect against reactive oxygen species. Cytotechnology 40: 139–149.
76. Ye J, Li Y, Hamasaki T, Nakamichi N, Komatsu T, et al. (2008) Inhibitory effect of electrolyzed reduced water on tumor angiogenesis. Biol Pharm Bull 31(1): 19–26. doi: 10.1248/bpb.31.19
77. Yan H, Kashiwaki T, Hamasaki T, Kinjo T, Teruya K, et al. (2011) The neuroprotective effects of electrolyzed reduced water and its model water containing molecular hydrogen and Pt nanoparticles. BMC Proc 5 (Suppl 8)69–70. doi: 10.1186/1753-6561-5-s8-p69
78. Kinjo T, Ye J, Yan H, Hamasaki T, Nakanishi H, et al. (2012) Suppressive effects of electrochemically reduced water on matrix metalloproteinase-2 activities and in vitro invasion of human fibrosarcoma HT1080 cells. Cytotechnology 64: 357–371. doi: 10.1007/s10616-012-9469-7
79. Teruya K, Myojin-Maekawa Y, Shimamoto F, Watanabe H, Nakamichi N, et al. (2013) Protective effects of the fermented milk kefir on X-ray irradiation-induced intestinal damage in B6C3F1 mice. Biol Pharm Bull 36(3): 352–359. doi: 10.1248/bpb.b12-00709
80. Vijayalaxmi, Reiter RJ, Tan D-X, Herman TS, Thomas CR (2004) Melatonin as a radioprotective agent: A review. Int J Radiat Oncol Biol Phys 59(3): 639–653. doi: 10.1016/j.ijrobp.2004.02.006
81. Akpolat M, Kanter M, Uzal MC (2009) Protective effects of curcumin against gamma radiation-induced ileal mucosal damage. Arch Toxicol 83: 609–617. doi: 10.1007/s00204-008-0352-4

Citations

CHAPTER 1

Jang Min Park, Dong-Wook Oh, and Jungho Lee, "Numerical Analysis of Thermal Mixing in a Swirler-Embedded Line-Heater for Flow Assurance in Subsea Pipelines," Advances in Mechanical Engineering, Article ID 739089, in press.

CHAPTER 2

Abhishek Joshi, Jitendra S Sangwai, Kousik Das, and Nagham Amer Sami, Experimental investigations on the phase equilibrium of semi-clathrate hydrates of carbon dioxide in TBAB with small amount of surfactant, doi:10.1186/2251-6832-4-11.

CHAPTER 3

E. Dendy Sloan, Carolyn A. Koh, and Amadeu K. Sum, Gas Hydrate Stability and Sampling: The Future as Related To the Phase Diagram, doi:10.3390/en3121991.

CHAPTER 4

Nakajima T, Kudo T, Ohmura R, Takeya S, Mori YH (2012) Molecular Storage of Ozone in a Clathrate Hydrate: An Attempt at Preserving Ozone at High Concentrations. PLoS ONE 7(11): e48563. doi:10.1371/journal.pone.0048563.

CHAPTER 5

Gordienko R, Ohno H, Singh VK, Jia Z, Ripmeester JA, et al. (2010) Towards a Green Hydrate Inhibitor: Imaging Antifreeze Proteins on Clathrates. PLoS ONE 5(2): e8953. doi:10.1371/journal.pone.0008953.

CHAPTER 6

Lau EY, Wong SE, Baker SE, Bearinger JP, Koziol L, et al. (2013) Comparison and Analysis of Zinc and Cobalt-Based Systems as Catalytic Entities for the Hydration of Carbon Dioxide. PLoS ONE 8(6): e66187. doi:10.1371/journal.pone.0066187.

CHAPTER 7

Lai Y-C, Chen Y-F, Chiang Y-W (2013) ESR Study of Interfacial Hydration Layers of Polypeptides in Water-Filled Nanochannels and in Vitrified Bulk Solvents. PLoS ONE 8(6): e68264. doi:10.1371/journal.pone.0068264.

CHAPTER 8

Hamasaki T, Nakamichi N, Teruya K, Shirahata S (2014) Removal Efficiency of Radioactive Cesium and Iodine Ions by a Flow-Type Apparatus Designed for Electrochemically Reduced Water Production. PLoS ONE 9(7): e102218. doi:10.1371/journal.pone.0102218.

Index

A
Antifreeze proteins (AFPs) 73, 74

C
Carbonic anhydrase (CA) 92
Carbonic anhydrase II (CAII) 91
Circular dichroism (CD) 130
Conductor-like polarizable continuum model (CPCM) 96

D
Direct electric heating (DEH) 3
Dissolved oxygen (DO) 75
Double electron-electron resonance (DEER) 127

E
Electrical heater 2, 4
Electrically trace heated pipe-in-pipe (ETH-PiP) 4
Electrochemically reduced water (ERW) 167
Electron nuclear double resonance (ENDOR) 144
Electron spin echo envelope modulation (ESEEM) 127
Electron spin echo (ESE) 127
Electron spin resonance (ESR) 127
Encounter complex (EC) 98, 99

F

Fourier Transformed (FT) 139
Fukushima Daiichi Nuclear Power Plant (FDNPP) 164

G

Gas hydrate occurrence zone (GHOZ) 44
Gas hydrate stability zone (GHSZ) 44
Green fluorescent protein (GFP) 75

H

Honestly significant difference (HSD) 85
Human carbonic anhydrase (HCAII) 100
Hydrate-bearing sediment (HBS) 51
Hydrate composition 48

L

Lolium perenne (Lp) 75

M

Methanethiosulfonate spin label (MTSL) 131

N

Natural gas 2

O

Oxidation-reduction potential (ORP) 170
Oxygen of water (OW) 103

P

Phase diagram 41, 42, 46, 48, 49, 50, 53
Pipe-in-pipe (PiP) 3
Powder X-ray diffraction (PXRD) 65

R

Reactive oxygen species (ROS) 166

S

Structure II (sII) 73, 74

T

Tetra hydro furan (THF) 74
Thermal hysteresis (TH) 75

U

United atom topological model (UA0) 96